# Imitating Humans

## A Technical Approach

**Siddharth Shrotriya**
**Animesh Pandey**

LULU PRESS

Imitating Humans: A Technical Approach
By Siddharth Shrotriya and Animesh Pandey

Published by:
Lulu Press,
Raleigh, NC
www.lulu.com

First Edition
January, 2013

ISBN 13:   978-1-300-65289-2

# Preface

The book which you are holding was written by two final year under-graduate students out of frustration on the existing ways of getting Computer Science education. We never have believed in rote learning and attention only being given to theory. This book's contents have been designed by for students who want to study Artificial Intelligence and Machine Learning in a more comprehendible form. In our under-graduation, we have seen students being taught this course in their colleges but due to lack of a practical approach these topics settle as theoretical stockpile in the students' minds. We also found that considerable amount of time was spent in making the student get familiarized with the simplest of topics as the focus was theoretical concepts rather than quick running applications. The brighter students eventually caught on to what was being done but the weaker students were not able to deal with it. Based on this experience, we reached on a firm conviction that the practical application must be given more importance than theoretical concepts as they are less time taken and mostly self-explanatory.

The text introduces many abstract concepts. The book is dealing with the human behavior, thinking and cognitive abilities and their mapping to the machine and computer world. A student when relates a subject to himself, a process of thinking develops within which makes him find answers himself. When he will try to think over the basic human thought and thought changes that occur in daily life, he might be able to model the possible solution of the computer science problem. This is where cognitive science intersects computer science. Equal emphasis has been given to both the sciences. One has thought of Artificial Intelligence as his own intelligence, the reason behind it and its use. In cognitive sciences, we have explained all topics related to the human behavior focusing mainly on the sensory characteristics.

There are diagrams and code snippets that will give you a hint on what are you actually studying. There are flow-charts or labeled diagrams for complex topics. We did not use traditionally accepted programming languages as this would increase the complexity of any topics by considerable amount. As a result we chose the programming languages of the best software companies - C Sharp (C#) from Microsoft and Go! from Google. C# is in widespread use, so it is also very much universally accepted. Go! is a new language but its advantages like fast compilation and concurrency support and garbage collection made us consider it. This will help getting more out of this language's potential and we were successful in doing that.

We have spent much of our time in trying to understand how this book can be of maximal benefit to the readers. We hope that it is!


SIDDHARTH SHROTRIYA

Jaypee Institute of Information Technology

Final Year B.Tech

Dept. of Electronics and Communication


ANIMESH PANDEY

Jaypee Institute of Information Technology

Final Year B.Tech

Dept. of Information Technology

# Table of Contents

# Chapter 1

# Human Behavior and Characteristics

"No one really knows why humans do what they do"

– Dr. David K. Reynolds

The quote was given by a prominent writer who has written many books on the characteristics of human behavior. This has been seen that it is very difficult to judge and predict the human behavior. It requires high amount of prior knowledge about the different type of behaviors, and one might not be enthusiastic enough to learn all of it. So, this may come to us as we grow older as by then we must have interacted with, perceived and judged a lot of people. On the basis of all the observations when we see a person we tend to guess the possible reasons behind his behavior.

This behavior thing is influenced by both extrinsic and intrinsic factors. "What we believe is heavily influenced by what we think others believe" — Thomas Gilovich, How We Know What Isn't So: The Fallibility of Human Reason in Everyday Life. We as humans are highly manipulative as well as highly perceptible to manipulation. We have identified five intrinsic factors which are relevant to this book's later chapters. Those factors are emotions, memory, learning like simple and complex learning, knowledge and perception.

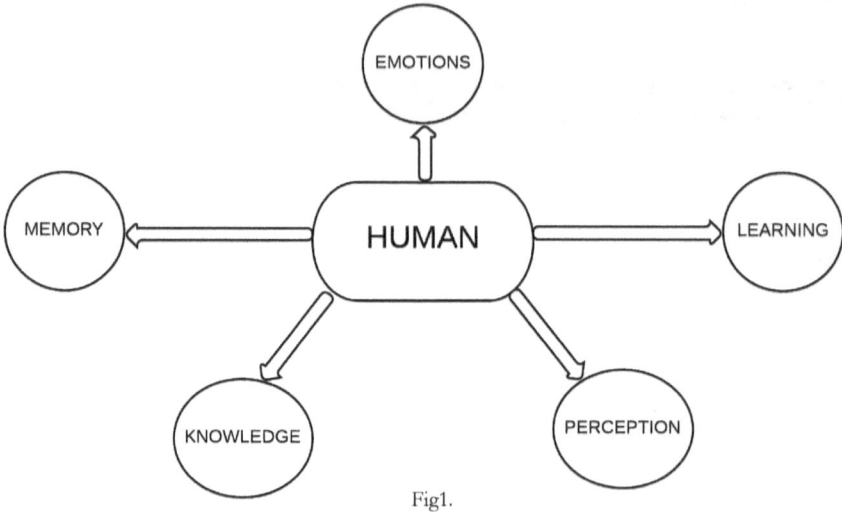

Fig1.

# Emotions

The nature is governed by response and stimulus. Emotions have been described as discrete and consistent responses to intrinsic or extrinsic events which have a particular significance for the human (Fig. 1). Emotions can be differentiated from a number of similar constructs within the field of affective neuroscience:

- Feelings are best understood as a subjective representation of emotions, private to the individual experiencing them.

- Moods are diffuse affective states that generally last for much longer durations than emotions and are also usually less intense than emotions.

- Affect is an encompassing term, used to describe the topics of emotion, feelings, and moods together, even though it is commonly used interchangeably with emotion.

Emotions play a central role in most forms of natural human interaction so we may expect that computational methods for the processing and expression of emotions will play a growing role in human-computer interaction. There are certain ways by which one can judge the emotion that one has:

- Cognitive appraisal: provides an evaluation of events and objects

- Bodily symptoms: the physiological component of emotional experience

- Action tendencies: a motivational component for the preparation and direction of motor responses.

- Expression: facial and vocal expression almost always accompanies an emotional state to communicate reaction and intention of actions

- Feelings: the subjective experience of emotional state once it has occurred

Ortony, Clore and Collins developed a computational emotion model, that is often referred to as the OCC model, which has established itself as the standard model for emotion synthesis. A large number of studies employed the OCC model to generate emotions. This model specifies 22 emotion categories based on valence reactions to situations constructed either as being goal relevant events, as acts of an accountable agent (including itself), or as attractive or unattractive objects. It also offers a structure for the variables, such as likelihood of an event or the familiarity of an object, which determines the intensity of the emotion types. It contains a sufficient level of complexity and detail to cover most situations an emotional interface character might have to deal with.

The OCC model has established itself as the standard model for emotion synthesis. A large number of studies employed the OCC model to generate emotions for their embodied characters. Many developers of such characters believe that the OCC model will be all they ever need to equip their character with emotions. This study reflects on the limitations of the OCC model specifically, and on the emotion models in general due to their dependency on artificial intelligence. OCC model has been connected to a very important aspect of Computer Science i.e. Artificial Intelligence which will be explained in later chapter.

Artificial characters need an emotion model to synthesize emotions and express them. The emotion model should enable the character to argue about emotions the way humans do. An event that upsets humans, for example the loss of money, should also upset the character. The emotion model must be able to evaluate all situations that the character might encounter and must also provide a structure for variables influencing the intensity of an emotion. Such an emotion model enables the character to show the right emotion with the right intensity at the right time, which is necessary for the convincingness of its emotional expressions.

The process of finding the resulting behavior of a human or artificial character on the basis of initial events i.e. Perception and New information can be split into three phases:

- Categorization - In the categorization phase the character evaluates an event, action or object, resulting in information on what emotional categories are affected.

- Quantification - In the quantification phase, the character calculates the intensities of the affected emotional categories.

- Mapping - The OCC model distinguishes 22 emotional categories. These need to be mapped to a possibly lower number of different emotional expressions.

# Memory

Memory refers to the processes that are used to obtain, store, retain and later retrieve that information when required. There are three main sub-processes involved in memory: encoding, storage and retrieval. In order to form new memories, information must be changed into a usable form, which occurs through the process known as encoding. Once information has been successfully encoded, it must be stored in memory for later use. Much of this stored memory lies outside of our awareness most of the time, except when we actually need to use it. The retrieval process allows us to bring stored memories into conscious awareness.

While several different models of memory have been proposed, the stage model of memory is often used to explain the basic structure and function of memory. Initially proposed in 1968 by Atkinson and Shiffrin, this theory outlines three separate stages of memory: sensory memory, short-term memory and long-term memory.

- **Sensory Memory**: Sensory memory is the earliest stage of memory. During this stage, sensory information from the environment is stored for a very brief period of time, generally for no longer than a half-second for visual information and 3 or 4 seconds for auditory information. We attend to only certain aspects of this sensory memory, allowing some of this information to pass into the next stage - short-term memory.

- **Short-Term Memory**: Short-term memory, also known as active memory, is the information we are currently aware of or thinking about. In Freudian psychology, this memory would be referred to as the conscious mind. Paying attention to sensory memories generates the information in short-term memory. Most of the information stored in active memory will be kept for approximately 20 to 30 seconds. While many of our short-term memories are quickly forgotten, attending to this information allows it to continue on the next stage - long-term memory.

- **Long-Term Memory**: Long-term memory refers to the continuing storage of information. In Freudian psychology, long-term memory would be call the preconscious and unconscious. This information is largely outside of our awareness, but can be called into working memory to be used when needed. Some of this information is fairly easy to recall, while other memories are much more difficult to access.

## The Organization of Memory

The ability to access and retrieve information from long-term memory allows us to actually use these memories to make decisions, interact with others and solve problems. But how is information organized in memory? The specific way information is organized in long-term memory is not well understood, but researchers do know that these memories are arranged in groups. Clustering is used to organize related information into groups. Information that is categorized becomes easier to remember and recall.

One way of thinking about memory organization is known as the semantic network model. This model suggests that certain triggers activate associated memories. A memory of a specific place might activate memories about related things that have occurred in that location. For example, thinking about a particular campus building might trigger memories of attending classes, studying and socializing with peers.

# Knowledge

Knowledge refers to the practical use of information. The information is something that can be shared and transported but this is not the case with knowledge. Knowledge primarily involves a personal experience with the information. E.g. a person reading the results of an experiment is getting the information about the project but the person who performed the experiment has the knowledge about it.

Fig. 2 DIKW Pyramid

# Learning

Learning can be defined as a process that depends on the lasting change in behavior potential that comes from experience. Learning highly depends on perception which we will be talking about in later paragraphs. There are several types of learning processes in the human mind that shape the overall personality. Human behavior psychology is a very complex topic. According to a few well-known psychologists "Myers-Briggs" and "Keirsey" there are about sixteen distinct personality types, which define our personality. The constructs that we have mentioned play a vital role in deciding the personality type.

There are several types of learning that can be related with humans:

- **Simple non-associative learning**: It refers to those instances in which the human's behavior toward a stimulus changes in the absence of any apparent associated stimulus or event (such as a reward or punishment).

- **Associative Learning**: Associative learning is the process by which an association between two stimuli or a behavior and a stimulus is learned.

- **Observational Learning**: The learning process most characteristic of humans is imitation; one's personal repetition of an observed behavior, such as a dance. It is like a child learning by seeing his parents do a particular task.

- **Episodic Learning**: Episodic learning is a change in behavior that occurs as a result of an event. For example, a fear of dogs that follows being bitten by a dog is episodic learning.

- **Rote Learning**: Rote learning is a technique which avoids understanding the inner complexities and inferences of the subject that is being learned and instead focuses on memorizing the material so that it can be recalled by the learner exactly the way it was read or heard.

- **Informal Learning**: Informal learning occurs through the experience of day-to-day situations (for example, one would learn to look ahead while walking because of the danger inherent in not paying attention to where one is going). It is learning from life, during a meal at table with parents, play, exploring, etc.

- **Formal Learning**: Formal learning is learning that takes place within a teacher-student relationship, such as in a school system.

- **Non-formal Learning**: Non-formal learning is organized learning outside the formal learning system. For example:

learning by coming together with people with similar interests and exchanging viewpoints, in clubs or in (international) youth organizations, workshops.

- **Tangential Learning**: Tangential learning is the process by which people will self-educate if a topic is exposed to them in a context that they already enjoy.

# Perception

Perception means to experience our surroundings. This involves both the recognition stimuli and actions as a response to these stimuli. One can take a moment to think about all the objects, people and animals that he has perceived on a daily basis. When we sense an object, there is an experience that is stored as information in our mind. All of these things help make up our conscious and allow us to interact with the people and objects around us. Perception includes five senses: touch, sight, sound, taste and smell. It also includes what is known as proprioception, a set of senses involving the ability to detect changes in body positions and movements. It also involves the cognitive processes required to process information, such as recognizing the face of a friend or detecting a familiar sound.

We have proposed a model on the behavioral characteristics of a human so that we can connect them to robots and other autonomous systems that can be of great use to the human race.

The model is a Stimuli-Response model. It comprises of the main five behavioral components of human beings i.e. Perception, information/knowledge, emotions, memory and learning. We will start with an example, suppose there is a man standing on a crossroad, he wishes to cross the road to get to the other side. Here perception is the key to his movement. This man once had an accident while he was crossing the road and that incident is making him hesitant in doing it. He perceives the task of road crossing to be harmful to him because of that one incident. Following are the main points about the model:

- Perception and external information are the main stimuli that make the human do a job or think.

- The human has memory where the long term memory plays a vital role in deciding the response.

- The memory has some prior knowledge about the surroundings which may or may not affect the perception about the situation.

- When the human has perceived or is in process of perception, learning take place in the mind where the new information integrated with the prior information and some possible options are created and the best option is the response.

- The response is terms of an action that depicts an emotion which can also be affected directly by the perception.

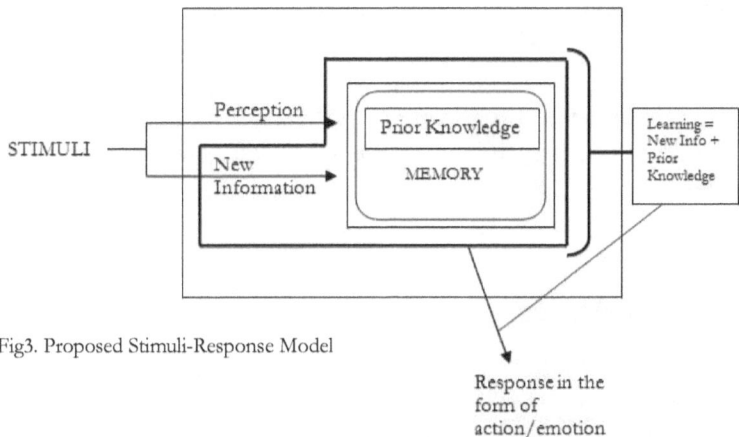

Fig3. Proposed Stimuli-Response Model

In the crossroads example, the perception is the roads and the vehicles around and external information may be that someone must have called him who is on the opposite end. He has a lot of memories in his mind but the accident is stored in his long term memory. His resultant emotion is fear and tension that comes out after learning that about the past accident under similar circumstances. Other fear could be of the fast cars that are moving around fearing that they might hit him is making him hesitant in crossing the road. So, this can be seen that every action is governed by some external and internal perceptions.

# Human Psychology in Problem Solving

A problem arises when we need to overcome some obstacle in order to get from our current state to a desired state. Problem solving is the process that an organism implements in order to try to get from the current state to the desired state.

## The Behaviorist Approach

Behaviorist researchers argued that problem solving was a "reproductive" process; that is, organisms faced with a problem applied behavior that had been successful on a previous occasion. Successful behavior was itself believed to have been arrived at through a process of trial-and-error. Suppose a human comes across a stray dog. To get past it, the human has to adopt certain measures which could be scaring the dog away with a stick or stone. He will adopt the way in which he was successful during the last encounter.

## The Gestalt Approach

By contrast, Gestalt psychologists argued that problem solving was a "productive" process. In particular, in the process of thinking about a problem individuals sometimes "restructured" their representation of the problem, leading to a flash of insight that enabled them to reach a solution. In The Mentality of Apes (1915) Wolfgang Köhler described a series of studies with apes in which the animals appeared to demonstrate insight in problem solving situations. A description of these studies, with photographs, can be found here.

## Problem Space Theory

In 1972, Allen Newell and Herbert Simon published the book *Human Problem Solving*, in which they outlined their problem space theory of problem solving. In this theory, people solve problems by searching in a problem space. The problem space consists of the initial (current) state, the goal state, and all possible states in between. The actions that people take in order to move from one state to another are known as **operators**.

There are three ways in which operators can be acquired:

- Trial-and-error. As noted above, this formed the basis of the behaviorist account of problem solving.

- Direct instruction.

- Analogies. Analogies are examples from one domain (the source), whose elements can be used to aid problem solving in another domain (the target).

Consider the "eight puzzle". The problem space for the eight puzzle consists of the initial arrangement of tiles, the desired arrangement of tiles (normally 1, 2, 3....8), and all the possible arrangements that can be arrived at in between. However, problem spaces can be very large so the key issue is how people navigate their way through the possibilities, given their limited working memory capacities.

For novel problems Newell and Simon proposed that operator selection is guided by cognitive short-cuts, known as **heuristics**. Heuristics relate to a usually speculative formulation serving as a guide in the investigation or solution of a problem. The simplest heuristic is repeat-state avoidance or backup avoidance1, whereby individuals prefer not to take an action that would take them back to a previous problem state. This is unhelpful when a person has taken an inappropriate action and actually needs to go back a step or more.

Another heuristic is "difference reduction, or hill-climbing", whereby people take the action that leads to the biggest similarity between current state and goal state.

A more sophisticated heuristic is "means-ends analysis". Like difference reduction, the means-ends analysis heuristic looks for the action that will lead to the greatest reduction in difference between the current state and goal state, but also specifies what to do if that action cannot be taken.

Some other problem-solving strategies that are used by humans are:

- **Algorithms:** An algorithm is a step-by-step procedure that will always produce a correct solution. A mathematical formula is a good example of a problem-solving algorithm. While an algorithm guarantees an accurate answer, it is not always the best approach to problem solving. This strategy is not practical for many situations because it can be so time-consuming.

  For example, if you were trying to figure out all of the possible number combinations to a lock using an algorithm, it would take a very long time!

- **Trial-and-Error**: A trial-and-error approach to problem-solving involves trying a number of different solutions and ruling out those that do not work. This approach can be a good option if you have a very limited number of options available. If there are many different choices, you are better off narrowing down the possible options using another problem-solving technique before attempting trial-and-error.

- **Insight**: In some cases, the solution to a problem can appear as a sudden insight. According to researchers, insight can occur because you realize that the problem is actually similar to something that you have dealt with in the past, but in most cases the underlying mental processes that lead to insight happen outside of awareness. The Behaviorist approach mentioned earlier is somewhat similar to this.

## Constraints and in Human way of Problem Solving

The human is not a perfect being. Distractions, fixations or wrong guesses can affect the efficiency of a solution. Some of the most common problems that arise are:

- Functional Fixedness: This term refers to the tendency to view problems only in their customary manner. Functional fixedness prevents people from fully seeing all of the different options that might be available to find a solution.

- Irrelevant or Misleading Information: When you are trying to solve a problem, it is important to distinguish between information that is relevant to the issue and irrelevant data that can lead to faulty solutions. When a problem is very complex, the easier it becomes to focus on misleading or irrelevant information.

- Assumptions: When dealing with a problem, people often make assumptions about the constraints and obstacles that prevent certain solutions.

- Mental Set: Another common problem-solving obstacle is known as a mental set, which is the tendency people have to only use solutions that have worked in the past rather than looking for alternative ideas. A mental set can often work as a heuristic, making it a useful problem-solving tool. However, mental sets can also lead to inflexibility, making it more difficult to find effective solutions.

Apart from the above mentioned problems, there is another problem called **Cognitive Bias** which is a pattern of deviation in judgment that occurs in particular situations, which may sometimes lead to perceptual distortion, inaccurate judgment, illogical interpretation, or what is broadly called irrationality.

This can be well explained by taking an example. Suppose you have to do a project. You have three constraints in front of you namely deadline, accuracy and originality. Sometimes you have to make up your mind or your mind makes you to focus on one thing more than the other. Too much attention on deadlines can highly affect the accuracy of the project, focus on accuracy may cause delay in completion and a lot of originality can also cause in delay.

Anchoring or focalism is one form of cognitive bias that describes the common human tendency to rely too heavily on the first piece of information offered (the "anchor") when making decisions. During decision making, anchoring occurs when individuals use an initial piece of information to make subsequent judgments. There are several factors influencing this bias:

- Mood: The person's current feelings and state of mind.

- Experience: It has been seen that people with expertise are also susceptible to anchoring.

- Personality: A person's personality i.e. how he sees the world, also plays a major role in his tendency of anchoring.

This brings us to the end of our first chapter about human behavior.

# Chapter 2

# Categorization of Human Cognition

"There can be no knowledge without emotion. We may be aware of a truth, yet until we have felt its force, it is not ours. To the cognition of the brain must be added the experience of the soul."

-Arnold Bennet

Cognition is a term referring to the mental processes involved in gaining knowledge and comprehension. These processes include thinking, knowing, remembering, judging, and problem-solving. These are higher-level functions of the brain and encompass language, imagination, perception, and planning.

Have you ever wondered why the human is so good in cognition? Episodic memory, non-linguistic mathematical ability, the capacity to navigate using landmarks, and our ability to make and use tools are all unique human traits. But they've all been documented behaviors in other animals. Some other peculiar abilities are:

- The ability to combine and recombine different types of information and knowledge in order to gain new understanding.

- Apply the same "rule" or solution to one problem to a different and new situation.

- To create and easily understand symbolic representations of computation and sensory input.

- To detach modes of thought from raw sensory and perceptual input.

In the paper *The cognitive niche: Co-evolution of intelligence, sociality, and language*, the author Steven Pinker poses Wallace's question anew:

". . . why do humans have the ability to pursue abstract intellectual feats such as science, mathematics, philosophy, and law, given that opportunities to exercise these talents did not exist in the foraging lifestyle in which humans evolved and would not have parlayed themselves into advantages in survival and reproduction even if they did?"

Pinker proposes an answer—that these feats are byproducts of selection for early humans to inhabit a "cognitive niche."

We will use a narrow interpretation of the word Cognition, which excludes sensory input analysis and motor control. The justification for this is that there is a fundamental difference between cognition and sensory input from evolutionary point of view. While sensory input analysis has evolved over long time to achieve the best performance in its tasks, cognition could not do that, because its tasks are too variable.

In sensory input analysis, it is reasonable to assume, even without any knowledge of the brain, that there are specialized systems to deal with specific tasks. it is not possible to predict the level of specialization, but we should not be surprised to find specialization even in very similar tasks. A specialization means a pattern of connections between neurons which is genetically defined, or at least is very constrained.

For example, recognition of horizontal and vertical movements are very similar tasks, but they still may be done by different systems. Both tasks were essential for human beings long before they became human, and stayed so during evolution, so specialized systems may evolve to execute them.

In cognition, the situation is different. Even if we look at very different cognitive tasks (e.g. debugging computer program vs making tea), they cannot be executed by specialized systems. This is because neither of this tasks has enough impact on human evolution to cause selection. This is true for almost all cognitive tasks (notable exception is recall, which is discussed below), and was so all through the latest stages of human development, because technical and cultural development move much faster than human biological evolution.

The conclusion is that the machinery behind cognition cannot be specialized to the task it is executing. Instead, the system must be a generic system, specialized to the task of learning new 'things', where things can be facts, associations, beliefs, techniques etc.

This is quite in contrast to the way many brain researchers think (e.g Churchland 1992, *The Computational Brain*, state specialization as the first feature of the brain). In most of the cases, they do this because they look mainly at sensory input analysis, and virtually ignore cognition in the sense that we use it here. In the discussion below we will emphasize the difference between genetically programmed features, which must be generic, and learned features, which can be specialized, but must be learned in some way.

We don't believe there is a sharp border between cognition and sensory input/motor control, and there probably are regions where specialized systems are mixed with parts of the generic system.

# Human Cognitive System

We will take for granted that Human Cognitive System (the System) (Fig1.) has the following physiological characteristics (genetically programmed):

- The System is made of neurons, mainly in the cortex.

- The actual operations are done by neurons activating/inhibiting other neurons (neuroglia cells may also take some part).

- The number of inputs and outputs of each neuron X is large, but much smaller than the total number of neurons N.

- The amount of information in the activation state of a single neuron is small, corresponding to only few bits.

- Some simple features are mediated by diffusing messengers in the extracellular and extra-synaptic environment.

- Learning (in any level) is done by changing the strength of the connections between neurons.

- The topology of connections between neurons (which connections exist, rather than their strength) in a mature System is essentially static over time.

- There is no way to access specific neurons by their address (which is the way data is accessed in computer).

- At any time, the number of active neurons is small compare the total number of neurons.

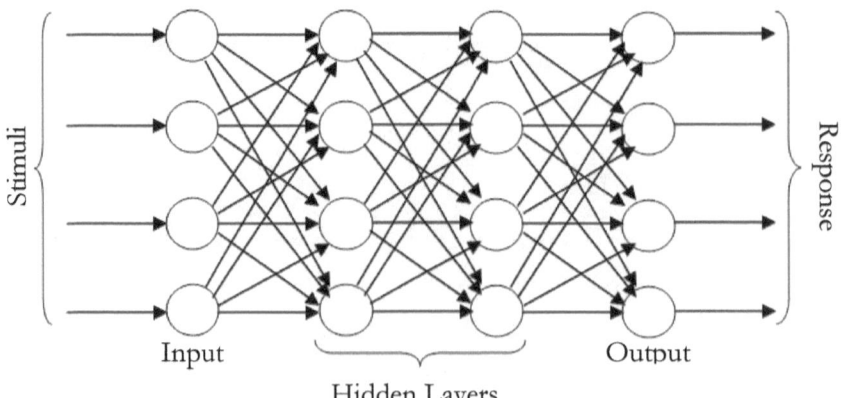

Fig1: Neural net for connecting stimuli to response

## Cognitive Features of the System

The Cognitive features that the System will hold are :

- The system deals effectively with large number of cognitive elements.

- The System can form association between virtually any pair of cognitive elements.

- The System is robust, and degrade gracefully. In general, cognitive elements are not associated with specific location in the System.

- The operations of the System are, to some extent, fuzzy, and it is not possible to repeat the same operation exactly.

- The System can learn new material.

- The System can learn to deal with complex situations, when the outcome is not obviously associated with actions that lead to it.

- The System can recall and manipulate information from its history. There is no limitation on the type of information that can be recalled. In particular, it can recall its own activity, though not in details.

- Episodic recall - the System tends to recall episodes from the near past (hours, up to few days). By Episode we mean groups of cognitive elements which had been in effect (in some sense of word) at the same time. Note that these include internal cognitive elements like concepts, intentions, emotions etc.

- Attention - The System seems to be limited in the number of cognitive elements it can deal with at the same time. The cognitive elements which are dealt with at any time are specified by one of:

  o The cognitive elements the system was dealing with until this time.

  o Some sensory input (in which we include things like pain, hunger etc.)

- Less frequently, the system deals with other elements. Mostly, these are elements to do with organization: Checking the time, checking if there is something the person need to do, etc.

- The person is aware of part of the operations of the System, but not all. Those elements which the person is aware of, he/she can directly recall later, by trying to recall his/her past (free recall).

- The System has a strong tendency to take actions which benefit the Person.

- The System has internal states. Some of these states tend to arise as a response to specific stimulus (or combination of stimuli, e.g. flower), and can lead to appropriate responses, but even in this case the association is not absolute. Other internal states are not associated with any specific stimulus.

- The System seems to be very good at pattern matching and associations, and less good at digital and Boolean operations.

## Micronodes and Cognitive Elements

A set of micronodes (set) is meaningless unless these micronodes are becoming active at the same time. If this happens only once, it is not really interesting. Thus only sets of micronodes which tend to become active together are interesting.

A set of micronodes will tend to become active together because of one (or both) of two reasons:

- They are all being activated by some another set of micronodes that are activated together. This maybe a result of sensory input, or as part of automatic operation (discussed below), or output of a global node.

- The set is self-sustaining. This means significant part of the micronodes in the set send output to other micronodes in the set. This kind of sets will tend to become active even if only part of their micronodes are activated from other nodes, and stay active for a long time.

# Episodic Recall and Awareness

By 'episodic Recall' we mean recall based on an episode in the history of the person, rather than on associations between cognitive elements. As discussed in the section about learning, this is essential part of the process of learning.

Since all we have is neurons, the recall of a cognitive element (== activation of a set of micronodes) must be mediated by some other neurons. Therefore, there must be some set of neurons which is activated when we try to do episodic recall. Since episodic recall is a specialized system, these neurons do not have to have the same characteristics as micronodes. For the discussion we will assume that they are, and that even if they don't, the main characteristics of the whole episodic recall system will still be the same.

The implementation of Episodic Recall System (ERS):

- The ERS is made of a large subset of the cognitive system ( 1--50 % of the total micronodes). All of these micronodes get input from a center, which is controlled by a smaller set. The center controls the level of activation in the ERS, and can also attenuate the activation signals from and to micronodes in the ERS. The inhibitory effect of the micronodes of the ERS on the rest of the System is very weak. The changes in strength of connections in the ERS are faster in both directions, i.e. both the increase and decrease in strength is faster.

- In normal operation, the center of the ERS keeps active relatively small number of micronodes (0.0001-0.01 of the total in the ERS). These micronodes do not have any specific pattern. This can be done by keeping the micronodes just below threshold level, and by attenuating down the activation into the ERS

- As a result, at any moment some of the micronodes in the ERS are active. These micronodes constitute the episode set. The strength of the connections between the episode set and the other active micronodes in the System is increased, and also the connections inside the episode set (feature 6 of micronodes). Since the micronodes which constitute the episode set are continuously changing, the episode set of any point in time is overlapping with both the preceding and following episode sets, and there are strengthen connection between the non-overlap micronodes of two neighboring episode-sets.

- When the System needs to do an episodic recall from the immediate past (e.g. Recall words from a list shortly after it was read), it does this by:
  - Attenuating the sensory input
  - Increasing activation in the ERS.

As a result, activation will spread from the current episode set to the preceding episode sets. This will send activation to the sets that were active at preceding time. If these matches the activation from the other cues (words, medium of input, etc.), these sets will become active (== the corresponding cognitive element will be recalled).

The activation in the ERS can continue to flow backwards, to recall more sets, but there is nothing that directs it to go only backwards in time. As a result, the activity becomes much diffused, and completely episodic recall cannot go much back in time. However, there are always additional cues active in the System, and the activation from these cues gives preference to the appropriate micronodes. This can be used by the System to prolong the episodic recall further in time.

## Evolution of the System

The mechanisms which are described here could evolve gradually, without sharp jumps. Each of the global nodes is useful on its own, so may have evolved independently of the rest of them. The learning mechanism, which involves cooperation between several systems, evolves on top of these. Thus, starting from a species with a central processor for sensory input and motor control, the cognition evolves in two 'macro' stages:

1. Evolution of global nodes and the ERS. It seems most likely that the main role of the global nodes in this stage is control of the alertness of the System. The ERS can evolve because it increases the ability of the System to recognize objects in the environment.

2. Evolution of the Learning Mechanism of operations. Once the ERS exists, this mainly involves connecting the pleasant sensory inputs and the *Pattern Match node* to the ERS. Each of these connections can evolve gradually and independently of the others. Having a learning mechanism is obviously beneficial at least in some circumstances, so once it started working it can evolve further.

Both of these stages can be gradual.

After the learning mechanism of operations evolves, the capabilities of the System are still limited by:

- Size

- Time to learn

- Rigidity of the input analysis and motor control systems. If these systems have genetically programmed wiring, they impose too much constrains on the development of the cognition system.

A small system would be restricted in the complexity of operations it can evolve. A large system can learn more, but it would take it more time to become effective in doing anything. It would also require the input and motor controls systems to co-develop with it. This explains why human babies are born so helpless: Their sensory-motor systems are 'intentionally' not complete, to allow them to co-develop with the cognition system.

Other animals probably have limited versions of the cognition system. It seems reasonable to assume that mammals in general have all the essential components of the learning mechanism, and the differences are quantitative, rather than qualitative. This is based mainly on the observation that ablation of the hippocampus, which presumably disrupt the ERS, gives similar effects in mice, monkeys and humans.

The requirement to have helpless, slow learning babies is probably the main obstacle in the evolution of complex cognition. The balance between the advantages of intelligence and this requirement determine whether a species will develop a high level intelligence or not (once the learning mechanism has evolved).

# Cognitive Categorization

Cognitive categorization theory suggests that humans have a tendency to label someone with descriptive words and phrases (first impression). Label people as soon as you meet them (friendly, funny, etc). People can't remember much about people besides how they were labeled (categorized). These labels are extremely resistant to change. This often happens during the job interview, with interviewer categorizing an interviewee within a couple of minutes of starting the interview.

In the last chapter it was mentioned that the Myers-Briggs and Kieresy types enlist 16 types of personalities. If known to a person, they can be a kind of cognitive qualities of some other person, when identified.

We organisms are sensorimotor systems. The things in the world come in contact with our sensory surfaces, and we interact with them based on what that sensorimotor contact "affords". All of our categories consist in ways we behave differently toward different kinds of things -- things we do or don't eat, mate-with, or flee-from, or the things that we describe, through our language, as prime numbers, affordances, or truths. That is all that cognition is for, and about.

Everything in nature is a dynamical system, of course, but some things are not only dynamical systems, and categorization refers to a special kind of dynamical system. Sand also interacts "differentially" with wind:

Blow it this way and it goes this way; blow it that way and it goes that way. But that is neither the right kind of systematicity nor the right kind of differentiality. It also isn't the right kind of adaptivity (though again, categorization theory probably has a lot to learn from ordinary dynamical interactions too, even though they do not count as categorization).

Few things that affect the categorization process are:

- **Learning**: When we see things in the nature we tend to relate them to one another. This helps us make clusters of objects and hence categorize them according to our convenience.

- **Innate Categories**: We as humans are different from each other. Our ways of perceiving things are also different from each other, which is the in born perception that we have. We sometimes tend to categorize things just in our mind.

- **Supervised Learning**: We sometimes bias our thoughts on the basis of previous results. We have tendency of going with the result that has a higher chance of occurring in the future.

- **Instrumental (operant/reinforcement) Learning**: For example, a pigeon is trained to peck at one key whenever it sees a black circle and at another key whenever it sees a white circle (with food as the feedback for doing the right thing and no-food as the feedback for doing the wrong thing) -- is already a primitive case of categorization.

  o It is a systematic differential response to different kinds of input, performed by an autonomous adaptive system that responded randomly at first, but learned to adapt its responses under the guidance of error-correcting feedback.

- **Color Categories**: When we see an image, we are able to differentiate between colors because of sudden change in color patterns.

- **Unsupervised Learning:** Such mechanisms cluster things according to their structural similarities and dissimilarities, enhancing both the similarities and the contrasts. An example of an unsupervised contrast-enhancing and boundary-finding mechanism is "reciprocal inhibition," in which activity from one point in visual space inhibits activity from surrounding points and vice-versa.

# Computer Science and Cognitive Science

In cognitive science, computers can be used in three ways:

- o   to simulate cognition for artificial intelligence,
- o   to enhance cognition by assisting human intellectual activity, and
- o   to help scientists understand cognition by testing theories on large amounts of data.

These three approaches are not mutually exclusive, since specialists in any of these areas frequently adopt techniques designed for the others. There are theories of categorization and reasoning in cognitive science that have been implemented and tested in computer systems. Most of the ideas originated long before modern computers were invented, but computers provide an opportunity for developing them in greater detail than was previously possible.

Theories of categorization in artificial intelligence, information retrieval, data mining, and other computational fields are no different in kind from theories that predate modern computers. The computer, however, introduces two important elements: it enables theories to be tested on large amounts of data, and it enforces precision, since no program running on a digital computer can ever be vague or ambiguous. Both elements can be helpful in formulating and testing theories, but neither can guarantee truth, relevance, or usefulness. Sometimes, as Lord Kelvin observed, precision can be a distraction: "Better a rough answer to the right question, than an exact answer to the wrong question." To avoid a bias toward answers that are easy to program, it is important to consider questions posed before computers were invented.

These computational methods can be used for three different purposes:

- **Artificial intelligence:** From the earliest days, computers were considered "giant brains", which had the potential to mimic and perhaps surpass human intelligence. Good (1965) predicted "It is more probable than not that, within the twentieth century, an ultra-intelligent machine will be built and that it will be the last invention that man need make." Except for chess playing, that prediction has not come to pass, but attempts to achieve it have contributed to a better appreciation of the richness and complexity of human intelligence.

- **Intelligence enhancement:** Computer capabilities are very different from and complementary to human abilities. That difference has led to a wide range of tools that supplement human cognition: information storage, management, search, query, and retrieval; text editing, analysis, translation, formatting, and distribution; calculation in graphics, spreadsheets, statistical packages, and formula manipulation; and computer-aided human communication and collaboration.

- **Hypothesis testing:** Psychology, linguistics, and philosophy deal with complex phenomena that cannot be described by the elegant mathematics used in physics. Without computers, theorists are often limited to studying an unrepresentative sample of oversimplified examples or to collecting statistics that show trends, but not causal connections. A computer, however, can analyze large volumes of realistic data and test hypotheses about causal mechanisms that could generate or interpret such data.

These three approaches differ primarily in their goals: simulation, enhancement, or understanding of human cognition.

Computational methods designed for any one of them can usually be adapted to the others.

## Categorization and Reasoning

Categorization and reasoning are interdependent cognitive processes: every multistep reasoning process depends on categorization at each step; every one-step reasoning process is also a one-step categorization process, and vice-versa; and every multistep categorization process involves multiple reasoning steps.

Three ways a person can come or can make a computer come to conclusion:

- **Deduction**: Apply a general principle to infer some fact.
  - Propositions: bird(x)->flies(x). bird(Tweety).

    Therefore, flies(Tweety).

    Categories: Birds⊂Flyers. Tweety∈Birds.

    Therefore, Tweety∈Flyers.

- **Induction**: Assume a general principle that explains many facts.
  - Propositions: bird(Tweety). bird(Polly). bird(Hooty). bat(Fred). flies(Tweety). flies(Polly). flies(Hooty). flies(Fred).

    Assume, bird(x)->flies(x).

    Categories: Birds={Tweety, Polly, Hooty}. Bats={Fred}.

    Flyers={Tweety, Polly, Hooty, Fred}.

    Assume, Birds⊂Flyers.

- **Abduction**: Guess a new fact that implies some given fact.
  - Propositions: bird(x)->flies(x). flies(Tweety).

    Guess, bird(Tweety).

Categories: Birds⊂Flyers. Tweety∈Flyers.

Guess, Tweety∈Birds.

These reasoning methods in Artificial Intelligence come under Knowledge Representation.


## Levels of Cognition

Whatever the neural mechanisms may be, the functions of human intelligence operate at multiple levels. The instant reaction to touching a hot stove is controlled by a wormlike ganglion well before the brain perceives what happened. Although muscles can be controlled by conscious attention, they operate most efficiently when the details are left to the fishlike parts of the brain.

In the first and this chapter we were introduced to various human behaviors and cognitive qualities.

The methods of studying and modeling cognition differ from one branch of cognitive science to another, but all of them are abstractions from nature. (Fig2.) The figure below depicts the evaluation of cognition. Level 1 is for microorganisms, level 2 can be for sea animals like fishes, level 3 is for cats/dogs and level4 is for mammals like humans.

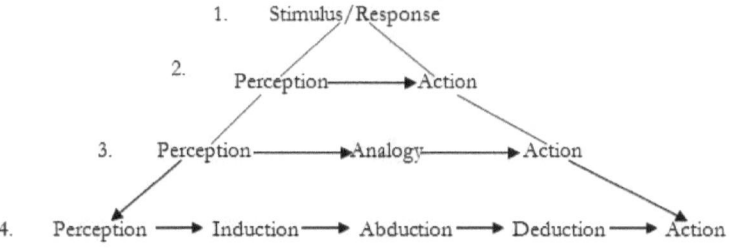

Fig2. Evaluation of Cognition


In summary, computer cognition gives us a better appreciation for the full range of human abilities. The greatest challenge for cognitive science is to integrate thousands of separate studies into a working model of the brain. In AI, Newell (1990) and Minsky (1987) integrated thirty or more years of their own research into models of cognition that have shown some promise. AI has been able to simulate most of the aspects of human intelligence and it has also contributed many valuable methods for enhancing intelligence and for testing hypotheses about intelligence.

This is the end of second chapter and we will discuss about the techniques of artificial intelligence in the next chapter.

# Chapter 3

# Using AI Techniques

## Introduction to Artificial Intelligence

Humankind has given itself the scientific name homo-sapiens because our mental capacities are so important to our everyday lives and our sense of self existence. The field of artificial intelligence, or AI, attempts to understand intelligent entities of human life. Thus, one reason to study it is to learn more about ourselves. But unlike philosophy and psychology, which are also concerned with intelligence, AI strives to build intelligent entities as well as understand them. Another reason to study AI is that these constructed intelligent entities are interesting and useful in their own right. AI has produced many significant and impressive products even at this early stage in its development. Although no one can predict the future in detail, it is clear that computers with human-level intelligence or better would have a huge impact on our everyday lives and on the future course of our civilization.

AI addresses most of the complex puzzles present in our life. How is it possible for a slow, tiny brain whether biological or electronic, to perceive, understand, predict, and manipulate a world far larger and more complicated than itself? How do we go about making something with those properties? These are hard questions, but unlike the search for faster-than-light travel or an antigravity device, the researcher in AI has solid evidence that the quest is possible. All the researchers have to do is look in the mirror to see an example of an intelligent system.

AI is one of the newest disciplines. It was formally initiated in 1956, when the name was coined, although at that point work had been under way for about five years. Along with modern genetics, it is regularly cited as the "field I would most like to be in" by scientists in other disciplines. A student in physics might reasonably feel that all the good ideas have already been taken by Galileo, Newton, Einstein, and the rest, and that it takes many years of study before one can contribute new ideas. AI, on the other hand, still has openings for a full-time Einstein.

The study of intelligence is also one of the oldest disciplines. For over 2000 years, philosophers have tried to understand how seeing, learning, remembering, and reasoning could, or should, be done. The advent of usable computers in the early 1950s turned the learned but armchair speculation concerning these mental faculties into a real experimental and theoretical discipline. Many felt that the new "Electronic Super-Brains" had unlimited potential for intelligence. "Faster than Einstein" was a typical headline for the AI system to be developed in the near future. AI has turned out to be more difficult than many at first imagined and modern ideas are much richer, more subtle, and more interesting as a result of the learning outputs of earlier systems.

AI currently encompasses a huge variety of subfields, from general-purpose areas such as perception and logical reasoning, to specific tasks such as playing chess, proving mathematical theorems, writing poetry and diagnosing diseases. Often, scientists in other fields move gradually into artificial intelligence, where they find the tools and vocabulary to systematize and automate the intellectual tasks on which they have been working all their lives. Similarly, workers in

AI can choose to apply their methods to any area of human intellectual endeavor. In this sense, it is truly a universal field.

If we want to state a formal definition of AI than it is the intelligence of machines and robots and the branch of computer science that aims to create it. AI textbooks define the field as "the study and design of intelligent agents" where an intelligent agent is a system that perceives its environment and takes actions that maximize its chances of success. John McCarthy, who coined the term in 1956, defines it as "the science and engineering of making intelligent machines."

AI research is highly technical and specialized, deeply divided into subfields that often fail to communicate with each other. Some of the division is due to social and cultural factors: subfields have grown up around particular institutions and the work of individual researchers. AI research is also divided by several technical issues. There are subfields which are focused on the solution of specific problems, on one of several possible approaches, on the use of widely differing tools and towards the accomplishment of particular applications. The central problems of AI include such traits as reasoning, knowledge, planning, learning, communication, perception and the ability to move and manipulate objects. General intelligence also referred to as strong AI is still among the field's long term goals. Currently popular approaches include statistical methods, computational intelligence and traditional symbolic AI. There are an enormous number of tools used in AI, including versions of search and mathematical optimization, logic, methods based on probability and economics, and many others. Although AI deals with a vast number of models and applications, there are four possible goals to pursue in artificial intelligence:

| Systems that think like humans | Systems that think rationally |
|---|---|
| Systems that act like humans | Systems that act rationally |

Historically, all four approaches have been followed. As one might expect, a tension exists between approaches centered on humans and approaches centered on rationality. We should point out that by distinguishing between human and rational behavior, we are not suggesting that humans are necessarily "irrational" in the sense of "emotionally unstable". One merely need note that we often make mistakes; we are not all chess grandmasters even though we may know all the rules of chess; and unfortunately, not everyone gets an A on the exam. A human-centered approach must be an empirical science, involving hypothesis and experimental confirmation. A rationalist approach involves a combination of mathematics and engineering. People in each group sometimes cast aspersions on work done in the other groups, but the truth is that each direction has yielded valuable insights. The different models universally linked to these approaches are shown below:

- **Acting humanly: The Turing Test approach**
- **Thinking humanly: The cognitive modeling approach**
- **Thinking rationally: The laws of thought approach**
- **Acting rationally: The rational agent approach**

As far as the scope of this chapter is concerned, we will mostly discuss about the acting humanly approach but the other models should be studied thoroughly to understand the amount of relativity between them and their applications.

**The Turing Test**, proposed by Alan Turing in 1950, was designed to provide a satisfactory operational definition of intelligence. Turing defined intelligent behavior as the ability to achieve human-level performance in all cognitive tasks, sufficient to fool an interrogator. Roughly speaking, the test he proposed is that the computer should be interrogated by a human via a teletype, and passes the test if the interrogator cannot tell if there is a computer or a human at the other end. For now, programming a computer to pass the test provides plenty to work on. The computer would need to possess the following capabilities:

- Natural language processing to enable it to communicate successfully in English (or some other human language);

- Knowledge representation to store information provided before or during the interrogation;

- Automated reasoning to use the stored information to answer questions and to draw new conclusions;

- Machine learning to adapt to new circumstances and to detect and extrapolate patterns.

Turing's test deliberately avoided direct physical interaction between the interrogator and the computer, because physical simulation of a person is unnecessary for intelligence. However, the so-called total Turing Test includes a video signal so that the interrogator can test the subject's perceptual abilities, as well as the opportunity for the interrogator to pass physical objects "through the hatch". To pass the total Turing Test, the computer will need

- Computer vision to perceive objects, and

- Robotics to move them about.

Within AI, there has not been a big effort to try to pass the Turing test. The issue of acting like a human comes up primarily when AI programs have to interact with people, as when an expert system explains how it came to its diagnosis, or a natural language processing system has a dialogue with a user. These programs must behave according to certain normal conventions of human interaction in order to make themselves understood. The underlying representation and reasoning in such a system may or may not be based on a human model. Now, we will discuss some of the techniques used in artificial intelligence to solve real life problems and systems trying to imitate humankind.

# Problem Solving

The term problem solving is used in many disciplines, sometimes with different perspectives, and often with different terminologies.

For instance, it is a mental process in psychology and a computerized process in computer science. The process of problem solving applied in human psychology has been discussed in the first chapter. Now, we will look into its version applied in computer science. In computer science and in the part of artificial intelligence that deals with algorithms, problem solving encompasses a number of techniques known as algorithms, heuristics, root cause analysis, etc. In these disciplines, problem solving is part of a larger process that encompasses problem determination, de-duplication, analysis, diagnosis, repair, etc.

Problem solving strategies are the steps that one would use to find the problems that in are in the way to getting to one's own goal. Some would refer to this as the 'problem-solving cycle'. In this cycle one will recognize the problem, define the problem, develop a strategy to fix the problem, organize the knowledge of the problem, figure-out the resources at the user's disposal, monitor one's progress, and evaluate the solution for accuracy. Although called a cycle, one does not have to do each step in order to fix the problem; in fact those who don't are usually better at problem solving. The reason it is called a cycle is that once one is completed with a problem another usually will pop up. Some part of complex problems consists of those who have only one solution like math problems or fact based questions which are grounded in psychometric intelligence. The other part is composed of those problems which are socio-emotional in nature and are unpredictable with answers that are constantly changing like what's your favorite color or what you should get someone for an occasion. The following techniques are usually called problem-solving strategies:

- Abstraction: solving the problem in a model of the system before applying it to the real system

- Brainstorming: suggesting a large number of solutions or ideas and combining and developing them until an optimum solution is found

- Divide and conquer: breaking down a large, complex problem into smaller, solvable problems

- Means-ends analysis: choosing an action at each step to move closer to the goal

- Method of focal objects: synthesizing seemingly non-matching characteristics of different objects into some new

- Reduction: transforming the problem into another problem for which solutions exist

- Research: employing existing ideas or adapting existing solutions to similar problems

- Trial-and-error: testing possible solutions until the right one is found

We will now discuss the implementation of some algorithms which come under these strategies. The algorithms relate to many strategies at the same time so it is difficult to segregate them separately. The most commonly used is search techniques which focus on finding an item with specified properties among a collection of items.


## Search Techniques

It is related to finding items from a collection items that may be stored individually as records in a database; or may be elements of a search space defined by a mathematical formula or procedure, such as the roots of an equation with integer variables; or a combination of the two, such as the Hamiltonian circuits of a graph. These techniques are usually applied for problems which occur when humans or machines have to make successive decisions whose outcomes are not entirely under one's control, such as in marketing, financial, military strategy planning or even in guidance of robots. This kind of problem is solved by combinatorial search which has been extensively studied in the context of artificial intelligence. Examples of algorithms for this class are the mini-max algorithm, alpha-beta pruning, and the A* algorithm.

## Uninformed Search Strategies

A problem determines the graph and the goal but not which path to select from the frontier. This is the job of a search strategy. A search strategy specifies which paths are selected from the frontier. Different strategies are obtained by modifying how the selection of paths in the frontier is implemented. This section presents those uninformed search strategies that do not take into account the location of the goal. Intuitively, these algorithms ignore where they are going until they find a goal and report success.

- **Depth first Search**: In this search, the frontier acts like a last-in first-out stack. The elements are added to the stack one at a time. The one selected and taken off the frontier at any time is the last element that was added. Implementing the frontier as a stack results in paths being pursued in a depth-first manner - searching one path to its completion before trying an alternative path. This method is said to involve backtracking: The algorithm selects a first alternative at each node, and it backtracks to the next alternative when it has pursued all of the paths from the first selection. Some paths may be infinite when the graph has cycles or infinitely many nodes, in which case a depth-first search may never stop. This algorithm does not specify the order in which the neighbors are added to the stack that represents the frontier. The efficiency of the algorithm is sensitive to this ordering.

- **Breadth first Search**: the frontier is implemented as a FIFO (first-in, first-out) queue. Thus, the path that is selected from the frontier is the one that was added earliest. This approach implies that the paths from the start node are generated in order of the number of arcs in the path. One of the paths with the fewest arcs is selected at each stage. Suppose the branching factor of the search is b. If the first path of the frontier contains n arcs, there are at least bn-1 elements of the frontier. All of these paths contain n or n+1 arcs. Thus, both space and time complexities are exponential in the number of arcs of the path to a goal with the fewest arcs. This method is guaranteed, however, to find a solution if one exists and will find a solution with the fewest arcs. Breadth-first search is useful when space is not a problem; you want to find the solution containing the fewest arcs; few solutions may exist, and at least one has a short path length; and infinite paths may exist, because it explores all of the search space, even with infinite paths. It is a poor method when all solutions have a long path length or there is some heuristic knowledge available. It is not used very often because of its space complexity.

- **Lowest cost first search:** When a non-unit cost is associated with arcs, we often want to find the solution that minimizes the total cost of the path. For example, for a delivery robot, costs may be distances and we may want a solution that gives the minimum total distance. Costs for a delivery robot may be resources required by the robot to carry out the action represented by the arc. The cost for a tutoring system may be the time and effort required by the students. In each of these cases, the searcher should try to minimize the total cost of the path found to reach the goal. The search algorithms considered thus far are not guaranteed to find the minimum-cost paths; they have not used the arc cost information at all. Breadth-first search finds a solution with the fewest arcs first, but the distribution of arc costs may be such that a path of fewest arcs is not one of minimal cost. The simplest search method that is guaranteed to find a minimum cost path is similar to breadth-first search; however, instead of expanding a path with the fewest number of arcs, it selects a path with the minimum cost. This is implemented by treating the frontier as a priority queue ordered by the cost function.

## Dijkstra's Algorithm

It is an algorithm for finding the shortest path between two graph vertices in a graph. It functions by constructing a shortest-path tree from the initial vertex to every other vertex in the graph. The worst-case running time for the Dijkstra algorithm on a graph with n nodes and m edges is $O(n^2)$ because it allows for directed cycles. It even finds the shortest paths from a source node s to all other nodes in the graph. This is basically $O(n^2)$ for node selection and $O(m)$ for distance updates. While $O(n^2)$ is the best possible complexity for dense graphs, the complexity can be improved significantly for sparse graphs. With slight modifications, Dijkstra's algorithm can be used as a reverse algorithm that maintains minimum spanning trees for the sink node. With further modifications, it can be extended to become bidirectional. Algorithm starts at the source vertex, s, it grows a tree, T, that ultimately spans all vertices reachable from S. Vertices are added to T in order of distance i.e., first S, then the vertex closest to S, then the next closest, and so on. Following implementation assumes that graph G is represented by adjacency lists.

Dijkstra (G, w, s)

Initialize single-source (G, s)

S ← { }        // S will ultimately contains vertices of final shortest-path weights from s

Initialize priority queue Q i.e., Q ← V[G]

while priority queue Q is not empty do

    u ← EXTRACT_MIN(Q)   // Pull out new vertex

    S ← S È {u}

    // Perform relaxation for each vertex v adjacent to u

    for each vertex v in Adj[u] do

        Relax (u, v, w)

Further implementation of this algorithm is shown by complete code demonstration in Go programming language as follows:

```go
package pathfinding

func minDist(dist map[*Node]int) *Node {
    var result_node *Node
    min := int(^uint(0) >> 1)
```

```
    for node, _dist := range dist {
        if min >= _dist {
            result_node = node
            min = _dist
        }}
    return result_node }

func removeFromQ(Q []*Node, n *Node) []*Node {
    var result []*Node
    for _, node := range Q {
        if node != n {
            result = append(result, node)
        }}
    return result  }

func dist_between(n1 *Node, n2 *Node) int {
    return 0  }

func Dijkstra(graph *Graph) []*Node {
    MAX_INT := int(^uint(0) >> 1)
    var path []*Node
    var Q []*Node
    dist := make(map[*Node]int, len(graph.nodes))

    for _, node := range graph.nodes {
        dist[node] = MAX_INT }
    dist[graph.start] = 0
    copy(graph.nodes, Q)
    for len(Q) != 0 {
        u := minDist(dist)
        if dist[u] == MAX_INT {
            break
        }
        Q = removeFromQ(Q, u)
        for _, v := range graph.adjacentNodes(u)
```

```
        { alt := dist[u] + dist_between(u, v)
              if alt < dist[u] {
                     dist[v] = alt
                     v.parent = u
                     //Reorder v in Q
              }}}
     return path
}
```

## A* search

It comes under the category of heuristic search algorithms. It is a computer algorithm that is widely used in path finding and graph traversal, the process of plotting an efficiently traversable path between points, called nodes. Noted for its performance and accuracy, it enjoys widespread use. This is the algorithm typically used in games like Warcraft-III. A* is not BFS (breadth first search), nor is it DFS (depth first search). In fact, it's a combination of Dijkstra's algorithm and best-first (greedy search). Don't worry, though, the algorithm is very easy to understand. The question arises while programming BFS and DFS: What if instead of blindly guessing the next node to traverse like in a simple DFS we pick the node which looks most promising? A* searches exactly like that: in a nutshell, we generate our possibilities and pick the one with the least projected cost. Once a possibility is generated and its cost is calculated, it stays in the list of possibilities until all the better nodes have been searched before it. First, let's define the cost function. The cost of a node, f, is given by:

$$f = g + h$$

In this, 'g' is the cost it took to get to the node, most likely the number of squares we passed by from the start. 'h' is our guess as to how much it will cost to reach the goal from that node. It's our heuristic (a heuristic, informally, is something which is not a definite series of steps to a solution (like an algorithm) but it helps us determine our answers is a rough way). A* will find the best path in a very short time provided the values of h is perfect. We don't have to recalculate g entirely each node. We can just add the distance to the parent node plus the parent's g. Many times, we can define h as the linear distance to the goal. This is best illustrated in a pseudo code:

```
// A* algorithm implemented using lists
initialize the open list
initialize the closed list
put the starting node on the open list (you can leave its f at zero)

while the open list is not empty
    find the node with the least f on the open list, call it "q"
    pop q off the open list
    generate q's 8 successors and set their parents to q
    for each successor
        if successor is the goal, stop the search
        successor.g = q.g + distance between successor and q
        successor.h = distance from goal to successor
        successor.f = successor.g + successor.h

        if a node with the same position as successor is in the open list
            which has a lower f than successor, skip this successor
        if a node with the same position as successor is in closed list
            which has a lower f than successor, skip this successor
        otherwise, add the node to the open list
    end
    push q on the closed list
end
```

It's important to note that we may search the same point a few times but only if the new path is more promising than the last time we searched it! Simply repeating mini-min search for each move ignores information from previous searches and results in infinite loops. In addition, since actions are committed based on limited information often the best move, may be due to undo the previous move. The principle of rationally is that backtracking should occur when the estimated cost of continuing the current path exceeds the cost of going back to a previous state plus the estimated cost of reaching the goal from the state Real-time A* (RTA*) implements the policy in constant time per move on a tree. For each move, the f(n) = g(n) + h(n) value of each neighbor of the current state is computed where is now the cost of the edge from the current state to the neighbor, instead of from the initial state. The problem solver moves to the neighbor with the minimum f(n) value, and stores with the previous state the best f(n) value among the remaining neighbors. This represents the h(n) value of the previous state from the perspective of the new current state. This is repeated until a goal is reached. To determine the h(n) value of a previously visited state, the stored value is used, while for a new state the heuristic evaluator is called. Note that the heuristic evaluator may employ minimum lookahead search with branch-and-bound as well.

In a problem space in which there exists a path to a goal from every state, RTA* is guaranteed to find a solution, regardless of the heuristic evaluation function. Moreover, on a tree, RTA* makes locally-optimal decisions given the information it has seen so far. To further explain the working of A* algorithm, a detailed code built in Go language is shown below to demonstrate its working in a real life-like environment solving a relatively complex map problem:

```
package pathfinding2
import (
    "fmt" )


//Defining possible graph elements
```

```go
const (
    UNKNOWN int = iota - 1
    LAND
    WALL
    START
    STOP )
type MapData [][]int

//Return a new MapData by value given some
dimensions
func NewMapData(rows, cols int) *MapData {
    result := make(MapData, rows)
    for i := 0; i < rows; i++ {
            result[i] = make([]int, cols)      }
    return &result   }

//A node is just a set of x, y coordinates with a
parent node and a
//heuristic value H
type Node struct {
    x, y    int //Using int for efficiency
    parent *Node
    H       int //Heuristic (aproximate distance)
    cost    int //Path cost for this node
}

//Create a new node
func NewNode(x, y int) *Node {
    node := &Node{
            x:       x,
            y:       y,
            parent: nil,
            H:       0,
            cost:    0,
    }
    return node
```

```go
}

//Return string representation of the node
func (self *Node) String() string {
    return fmt.Sprintf("<Node x:%d y:%d addr:%d>",
self.x, self.y, &self)
}

//Start, end nodes and a slice of nodes
type Graph struct {
    start, stop *Node
    nodes       []*Node
    data        *MapData
}

//Return a Graph from a map of coordinates (those
that are passible)
func NewGraph(map_data *MapData) *Graph {
    var start, stop *Node
    var nodes []*Node
    for i, row := range *map_data {
            for j, _type := range row {
                    if _type == START || _type == STOP
    {
                            node := NewNode(i, j)
                            nodes = append(nodes, node)
                            if _type == START {
                                    start = node }
                            if _type == STOP {
                                    stop = node  } } } }
    g := &Graph{
            nodes: nodes,
            start: start,
            stop:  stop,
            data:  map_data, }
    return g }
```

```
//Get *Node based on x, y coordinates.
func (self *Graph) Node(x, y int) *Node {
    //Check if node is not already in the graph
and append that node
    for _, n := range self.nodes {
            if n.x == x && n.y == y {
                return n
            } }
    map_data := *self.data
    if map_data[x][y] == LAND || map_data[x][y] ==
STOP {
            //Create a new node and add it to the
graph
            n := NewNode(x, y)
            self.nodes = append(self.nodes, n)
            return n      }
    return nil  }


//Get the nodes near some node
func (self *Graph) adjacentNodes(node *Node)
[]*Node {
    var result []*Node
    map_data := *self.data
    rows := len(map_data)
    cols := len(map_data[0])

//If the coordinates are passable then create a
new node and add it
if node.x <= rows && node.y+1 < cols {
if new_node := self.Node(node.x, node.y+1);
new_node != nil {
                result = append(result, new_node)
        }}
    if node.x <= rows && node.y-1 >= 0 {
            new_node := self.Node(node.x, node.y-1)
            if new_node != nil {
```

```
                    result = append(result, new_node)
        }}
    if node.y <= cols && node.x+1 < rows {
        new_node := self.Node(node.x+1, node.y)
        if new_node != nil {
            result = append(result, new_node)
        }}
    if node.y <= cols && node.x-1 >= 0 {
        new_node := self.Node(node.x-1, node.y)
        if new_node != nil {
            result = append(result, new_node)
        }}
    return result  }


func abs(x int) int {
    if x < 0 {
            return -x}
    return x }


func removeNode(nodes []*Node, node *Node) []*Node
{
    ith := -1
    for i, n := range nodes {
        if n == node {
            ith = i
            break
        }}
    if ith != -1 {
        copy(nodes[ith:], nodes[ith+1:])
        nodes = nodes[:len(nodes)-1]
    }       return nodes   }


func hasNode(nodes []*Node, node *Node) bool {
    for _, n := range nodes {
        if n == node {
            return true
```

```
        }}
    return false  }


//Return the node with the minimum H
func minH(nodes []*Node) *Node {
    if len(nodes) == 0 {
        return nil
    }
    result_node := nodes[0]
    minH := result_node.H
    for _, node := range nodes {
        if node.H < minH {
            minH = node.H
            result_node = node
        }}
    return result_node    }


func retracePath(current_node *Node) []*Node {
    var path []*Node
    path = append(path, current_node)
    for current_node.parent != nil {
        path = append(path, current_node.parent)
        current_node = current_node.parent }
    //Reverse path
    for i, j := 0, len(path)-1; i < j; i, j = i+1,
j-1 {
        path[i], path[j] = path[j], path[i]
    }
    return path }


// In our particular case: Manhatan distance
func Heuristic(graph *Graph, tile *Node) int {
    return abs(graph.stop.x-tile.x) +
abs(graph.stop.y-tile.y) }


func Astar(graph *Graph) []*Node {
```

```
    var path, openSet, closedSet []*Node
    openSet = append(openSet, graph.start)
    for len(openSet) != 0 {
            //Get the node with the min H
            current := minH(openSet)
            if current.parent != nil {
                    current.cost = current.parent.cost
+ 1         }
            if current == graph.stop {
                    return retracePath(current)   }
            openSet = removeNode(openSet, current)
            closedSet = append(closedSet, current)
            for _, tile := range
graph.adjacentNodes(current) {
        if tile != nil && graph.stop != nil &&
!hasNode(closedSet, tile) {
                    tile.H = Heuristic(graph, tile) +
current.cost

                    if !hasNode(openSet, tile) {
                    openSet = append(openSet,
tile) }
            tile.parent = current }}}
    return path
}
```

This code clearly shows the implementation of every part of the A* algorithm and enables us to use it in the real world to solve complex search problems and thus makes it one of the most important problem solving algorithm in artificial intelligence.

## Hill Climbing

In hill climbing the basic idea is to always head towards a state which is better than the current one. So, if you are at town A and you can get to town B and town C (and your target is town D) then you should make a move IF town B or C appear nearer to town D than town A does. In steepest ascent hill climbing you will always make your next state the best successor of your current state, and will only make a move if that successor is better than your current state. This can be described as follows:

a. Start with current-state = initial-state.

b. Until current-state = goal-state OR there is no change in current-state do:

    1. Get the successors of the current state and use the evaluation function to assign a score to each successor.

    2. If one of the successors has a better score than the current-state then set the new current-state to be the successor with the best score.

Note that the algorithm does not attempt to exhaustively try every node and path, so no node list or agenda is maintained - just the current state. If there are loops in the search space then using hill climbing you shouldn't encounter them - you can't keep going up and still get back to where you were before. Hill climbing terminates when there are no successors of the current state which are better than the current state itself.

## Simulated Annealing

There are certain optimization problems that become unmanageable using combinatorial methods as the number of objects becomes large. A typical example is the traveling salesman problem, which belongs to the NP-complete class of problems. For these problems, there is a very effective practical algorithm called simulated annealing (thus named because it mimics the process undergone by misplaced atoms in a metal when its heated and then slowly cooled). While this technique is unlikely to find the optimum solution, it can often find a very good solution, even in the presence of noisy data. The traveling salesman problem can be used as an example application of simulated annealing. In this problem, a salesman must visit some large number of cities while minimizing the total mileage traveled. If the salesman starts with a random itinerary, he can then pairwise trade the order of visits to cities, hoping to reduce the mileage with each exchange. The difficulty with this approach is that while it rapidly finds a local minimum, it cannot get from there to the global minimum.

Simulated annealing improves this strategy through the introduction of two tricks. The first is the so-called "Metropolis algorithm" in which some trades that do not lower the mileage are accepted when they serve to allow the solver to "explore" more of the possible space of solutions. Such "bad" trades are allowed using the criterion that

$$e^{-\Delta D/T} > R(0,1)$$

where $\Delta D$ is the change of distance implied by the trade (negative for a "good" trade; positive for a "bad" trade), $T$ is a "synthetic temperature," and $R(0,1)$ is a random number in the interval [0,1]. $D$ is called a "cost function," and corresponds to the free energy in the case of annealing a metal (in which case the temperature parameter would actually be the $k\,T$, where $k$ is Boltzmann's Constant and $T$ is the physical temperature, in the Kelvin absolute temperature scale). If $T$ is large, many "bad" trades are accepted, and a large part of solution space is accessed. Objects to be traded are generally chosen randomly, though more sophisticated techniques can be used.

The second trick is, again by analogy with annealing of a metal, to lower the "temperature." After making many trades and observing that the cost function declines only slowly, one lowers the temperature, and thus limits the size of allowed "bad" trades. After lowering the temperature several times to a low value, one may then "quench" the process by accepting only "good" trades in order to find the local minimum of the cost function. There are various "annealing schedules" for lowering the temperature, but the results are generally not very sensitive to the details.

There is another faster strategy called threshold acceptance (Dueck and Scheuer 1990). In this strategy, all good trades are accepted, as are any bad trades that raise the cost function by less than a fixed threshold. The threshold is then periodically lowered, just as the temperature is lowered in annealing. This eliminates exponentiation and random number generation in the Boltzmann criterion. As a result, this approach can be faster in computer simulations.

## Genetic Algorithms

Genetic algorithms were formally introduced in the United States in the 1970s by John Holland at University of Michigan. The continuing price/performance improvements of computational systems have made them attractive for some types of optimization. In particular, genetic algorithms work very well on mixed (continuous and discrete), combinatorial problems. They are less susceptible to getting 'stuck' at local optima than gradient search methods. But they tend to be computationally expensive in terms of execution of the algorithm.

To use a genetic algorithm, you must represent a solution to your problem as a genome (or chromosome). The genetic algorithm then creates a population of solutions and applies genetic operators such as mutation and crossover to evolve the solutions in order to find the best one(s). Now let take a look on the basic concept of this algorithm in the form of detailed explanation along with sample data and relevant pseudo codes.

Assume we have a discrete search space and a function:

$$\mathbf{f} : \varkappa \rightarrow \mathbf{R}$$

The general problem is to find

$$\min{}_{x \in \varkappa} (\mathbf{f})$$

Here $\mathbf{x}$ is a vector of decision variables, and $\mathbf{f}$ is the objective function. We assume here that the problem is one of minimization, but the modifications necessary for a maximization problem are nearly always obvious. Such a problem is commonly called discrete or combinatorial optimization problems (COP). One of the distinctive features of the GA approach is to allow the separation of the representation of the problem from the actual variables in which it was originally formulated. In line with biological usage of the terms, it has become customary to distinguish the 'genotype'—the encoded representation of the variables, from the 'phenotype'—the set of variables themselves. That is, the vector **x** is represented by a string **s**, of length l, made up of symbols drawn from an alphabet **A**, using a mapping $c : A' \rightarrow x$.

In practice, we may need to use a search space **S** of which **A'** is a subset to reflect the fact that some strings in the image of **A'** under c may represent invalid solutions to the original problem. (This is a potential source of difficulty for GAs in combinatorial optimization) The string length l depends on the dimensions of both $x$ and **A** and the elements of the string correspond to 'genes', and the values those genes can take to 'alleles'. This is often designated as the genotype–phenotype mapping. Thus the optimization problem becomes one of finding

$$\min{}_{s \,\in\, S} \, g(s)$$

where the function $g(s) = f(c(s))$. It is usually desirable that c should be a bijection. (The important property of a bijection is that it has an inverse, i.e., there is a unique vector **x** for every string **s**, and a unique string s for every vector x.) In some cases the nature of this mapping itself creates difficulties for a GA in solving optimization problems. In using this device, Holland's ideas are clearly distinct from the similar methodology developed by Rechenberg and Schwefel, who preferred to work with the original decision variables directly. Both Holland's and Goldberg's books claim that representing the variables by binary strings (i.e., A = {0,1}) is in some sense 'optimal', and although this idea has been challenged, it is still often convenient from a mathematical standpoint to consider the binary case. Certainly, much of the theoretical work in GAs tends to make this assumption. In applications, many representations are possible.

The original motivation for the GA approach was a biological analogy. In the selective breeding of plants or animals, for example, offspring are sought that have certain desirable characteristics—characteristics that are determined at the genetic level by the way the parents' chromosomes combine. In the case of GAs, a population of strings is used, and these strings are often referred to in the GA literature as chromosomes. The recombination of strings is carried out using simple analogies of genetic crossover and mutation, and the search is guided by the results of evaluating the objective function **f** for each string in the population. Based on this evaluation, strings that have higher fitness (i.e., represent better solutions) can be identified, and these are given more opportunity to breed. It is also relevant to point out here that fitness is not necessarily to be identified simply with the composition f(c(s)); more generally, fitness is h(f(c(s))) where **h : R → R** is a monotonic function.

Perhaps the most fundamental characteristic of genetic algorithms is that their use of populations of many strings. Here again, the German school of ES initially did not use populations, and focused almost exclusively on 'mutation' operators which are generally closer in concept to the types of operator used in neighborhood search and its extensions. Holland also used mutation, but in his scheme it is generally treated as subordinate to crossover. Thus, in Holland's GA, instead of the search moving from point to point as in NS approaches, the whole set of strings undergoes 'reproduction' in order to generate a new population. DeJong's work established that population-based GAs using crossover and mutation operators could successfully deal with optimization problems of several different types, and in the years since this work was published, the application of GAs to COPs has grown almost exponentially. At this point, however, it might be helpful to provide a very basic introduction. Crossover is a matter of replacing some of the genes in one parent by corresponding genes of the other. An example of one-point crossover would be the following. Given the parents P1 and P2, with crossover point 3 (indicated by X), the offspring will be the pair 01 and 02:

```
P1  1  0  1  0  0  1  0     01  1  0  1  1  0  0  1
             X
P2  0  1  1  1  0  0  1     02  0  1  1  0  0  1  0
```

The other common operator is mutation, in which a subset of genes is chosen randomly and the allele value of the chosen genes is changed. In the case of binary strings, this simply means complementing the chosen bits. For example, the string 01 above, with genes 3 and 5 mutated, would become 1 0 0 1 1 0 1. A simple pseudo code for the genetic algorithm is shown below:

> Choose an initial population of chromosomes;
> while termination condition not satisfied do
>   repaeat
>       if crossover condition satisfied then
>       {select parent chromosomes;
>       choose crossover parameters;
>       perform crossover};
>       if mutation condition satisifed then
>       {choose mutation points;
>       Perform mutation};
>       Evaluate fitness of offspring
>     untill sufficient offspring created;
>   select new population;
>   endwhile

To further understand the concept of this algorithm, a complete code implementation built in Go language is shown below:

```
//Code for genetic algorithm
  package genetic

  import (
      "fmt";
      "rand";
      "container/list";
      "time"; )
```

```go
// algorithm configuration
const (
    ProblemNumBits = 64;
    ProbCrossover = 0.98; // [0,1]
    ProbMutation = 1.0/ProblemNumBits; // [0,1]
    SelectionNumBouts = 3; // <=PopulationSize
    PopulationSize = 50; // expect even
    AlgorithmNumGenerations = 50; )

type Solution struct {
    bitstring []bool;
    score float; }

func NewSolution() *Solution {
    s := new(Solution);
    s.bitstring = make([]bool, ProblemNumBits);
    s.score = 0;
    return s;                        }

func NewRandomSolution() *Solution {
    s := NewSolution();
    // populate with random values
    for i:=0; i<len(s.bitstring); i++ {
        // rand in [0,1]
        if rand.Intn(2) == 1 {
            s.bitstring[i] = true;
        }    }
    return s;    }

func (s *Solution) String() string {
    return fmt.Sprintf("(%v) [%v]", s.score,
BitstringToString(&s.bitstring));  }

func BitstringToString(bitstring *[]bool) (s
string) {
```

```
    for i:=0; i<len(*bitstring); i++ {
        // no ternary
        if (*bitstring)[i] {
            s+="1";
        }else{
            s+="0";
        }    }
    return s;    }


func OneMax(bitstring *[]bool) (score float) {
    for _, v := range *bitstring {
        if v { score += 1.0; }
            }
    return score;    }


func Fitness(s *Solution) {
    s.score = OneMax(&s.bitstring);    }


func InitializePopulation() (*list.List) {
    population := list.New();

    c := make(chan *Solution, PopulationSize);

// create all of the population in parallel
    for i:=0; i<PopulationSize; i++ {
        go func() {c<-NewRandomSolution();}();
    }
    // barrier to collect results
    for i:=0; i<PopulationSize; i++ {
        population.PushBack(<-c);
    }
    return population; }


func EvaluatePopulation(population *list.List)
(best *Solution) {
    c := make(chan *Solution, population.Len());
```

```go
// evaluate the population in parallel
    for e := population.Front(); e != nil; e =
e.Next() {
        var s = e.Value.(*Solution);
        go func() {Fitness(s);c<-s}();
    }
    // barrier waiting for all evaluations to
finish, locate best result
    for i:=0; i<population.Len(); i++ {
        var s = <-c;
        if best==nil || s.score>best.score {
            best = s;
        }       }
    return best; }


func Mutate(candidate *Solution) {
    for i:=0; i<len(candidate.bitstring); i++ {
        if rand.Float32() < ProbMutation {
            candidate.bitstring[i] =
!candidate.bitstring[i]; // bit flip
        } } }


func Crossover(parent1 *Solution, parent2
*Solution) (*Solution, *Solution) {
    child1, child2 := NewSolution(),
NewSolution();
    // crossover point
    var pt = rand.Intn(len(parent1.bitstring));
    useCrossover := true;
if rand.Float32()>ProbCrossover {useCrossover =
false;}
    // can we use slices?
    for i:=0; i<len(parent1.bitstring); i++ {
        if useCrossover && i<pt {

    child1.bitstring[i],child2.bitstring[i] =
parent1.bitstring[i],parent2.bitstring[i];
```

```
        } else {

    child1.bitstring[i],child2.bitstring[i] =
parent2.bitstring[i],parent1.bitstring[i];
        }     }
    return child1, child2; }


func MakeChildren(parent1 *Solution, parent2
*Solution, channel chan *Solution) {

    // crossover
    child1, child2 := Crossover(parent1, parent2);
    // perform mutation
    Mutate(child1);
    Mutate(child2);
    // push out results
    channel<-child1;
    channel<-child2;
}


func Reproduce(population *list.List) (*list.List)
{
    children := list.New();
    channel := make(chan *Solution,
population.Len()); // with buffer

    // evaluate the population in parallel
    for e1:=population.Front(); e1 != nil;
e1=e1.Next() {
        e2 := e1.Next(); // expects even pop
size
        // reproduce in parallel
        go MakeChildren(e1.Value.(*Solution),
e2.Value.(*Solution), channel);
        // ensure we end safely
        e1 = e2;
    }
    // barrier to collect results
```

```
    for i:=0; i<population.Len(); i++ {
        children.PushBack(<-channel);
    }
    return children;
}


func Select(population *list.List) (*list.List) {
    selected := list.New();
    channel := make(chan *Solution,
population.Len()); // with buffer
    // select
    for i:=0; i<population.Len(); i++ {
        go func() {
            var best *Solution;
            for j:=0; j<SelectionNumBouts; j++
{
        // pick
        index := rand.Intn(population.Len());
      var candidate *Solution; var k = 0;
for e := population.Front(); e != nil; e =
e.Next() {
    if(k==index) {
    candidate = e.Value.(*Solution);
        break; }
        k++;   }
    // test
    if(best==nil || candidate.score>best.score) {
best = candidate; }
            }
            channel<-best;
        }();
    }
    // barrier to collect results
    for i:=0; i<population.Len(); i++ {
        selected.PushBack(<-channel);
    }
    return selected;
```

```
}

func Evolve() (best *Solution){
    // initialize the population
    population := InitializePopulation();
    // evolve the population in a sequence
    for i:=0; i<AlgorithmNumGenerations; i++ {
            // evaluate
        var s = EvaluatePopulation(population);
if best==nil || s.score>best.score { best = s;}
 fmt.Printf(" > gen %v, best=%v\n", i, best);
            // select
            selected := Select(population);
            // reproduce
            children := Reproduce(selected);
            // replace
            population = children;
    }
    return best;
}


// entry point
func main() {
// set the random seed
    seed := time.LocalTime().Seconds();
    rand.Seed(seed);
// run the optimization and get the best solution
    fmt.Printf("Starting...[seed=%v]\n", seed);
    best := Evolve();
// display the result
    fmt.Printf("Finished!\nThe best solution is:
%v\n", best);
}
```

The code is self-explanatory and is implemented using the main function declared at the end of the code. To run this code or any other code in your computer, you have to download the Go compliers from its website and then test run these algorithms.

## Minimax Algorithm

It is an example of adversarial search which deals with problems where two or more parties are competing against each other. Minimax is a decision rule used in decision theory, game theory, statistics and philosophy for minimizing the possible loss for a worst case scenario. Alternatively, it can be thought of as maximizing the minimum gain (maxi-min). Originally formulated for two-player zero-sum game theory, covering both the cases where players take alternate moves and those where they make simultaneous moves, it has also been extended to more complex games and to general decision making in the presence of uncertainty. The mini-max theorem states for every two-person, zero-sum game with finitely many strategies, there exists a value V and a mixed strategy for each player, such that

(a) Given player 2's strategy, the best payoff possible for player 1 is V, and

(b) Given player 1's strategy, the best payoff possible for player 2 is −V.

Equivalently, Player 1's strategy guarantees him a payoff of V regardless of Player 2's strategy, and similarly Player 2 can guarantee himself a payoff of −V. The name mini-max arises because each player minimizes the maximum payoff possible for the other—since the game is zero-sum, he also minimizes his own maximum loss (i.e. maximize his minimum payoff). This algorithm can be better understood by implementing in the form of code and arriving to best solution with the help of it. Here, the complete algorithm is demonstrated in C# language to show the correct procedure of evaluating this algorithm. The code describes the playing strategy of a tic-tac-toe game using minimax algorithm:

```
namespace ttt_mm
{
```

```
class AI
{
    // Implementation of the minimax algorithm.
Determines the best move for the current
    // board by playing every move combination
until the end of the game.
    public static Space GetBestMove(GameBoard gb,
Player p)
    {
        Space? bestSpace = null;
        List<space> openSpaces = gb.OpenSquares;
        GameBoard newBoard;

        for (int i = 0; i < openSpaces.Count; i++)
        {
            newBoard = gb.Clone();
            Space newSpace = openSpaces[i];
            newBoard[newSpace.X, newSpace.Y] = p;

            if (newBoard.Winner == Player.Open &&
newBoard.OpenSquares.Count > 0)
            {
                Space tempMove = GetBestMove(newBoard,
((Player)(-(int)p)));   //a little hacky, inverts
the current player
                newSpace.Rank = tempMove.Rank;
            }
            else
            {
                if (newBoard.Winner == Player.Open)
                    newSpace.Rank = 0;
                else if (newBoard.Winner == Player.X)
                    newSpace.Rank = -1;
                else if (newBoard.Winner == Player.O)
                    newSpace.Rank = 1;
            }
```

```
        //If the new move is better than our
previous move, take it
        if (bestSpace == null ||
            (p == Player.X && newSpace.Rank <
((Space)bestSpace).Rank) ||
            (p == Player.O && newSpace.Rank >
((Space)bestSpace).Rank))
        {
            bestSpace = newSpace;
        }
    }
    return (Space)bestSpace;
    }
  }
}
```

In this code, the main part is implemented by Getbestmove()
function which describes the working of the algorithm in order
to benefit the user to win this game and minimizes the chances of
the opposite player as stated in the algorithm.

## Alpha beta Pruning

Alpha–beta pruning is a search algorithm that seeks to decrease the
number of nodes that are evaluated by the minimax algorithm in its
search tree. It is an adversarial search algorithm used commonly for
machine playing of two-player games (Tic-tac-toe, Chess, etc.). It
stops completely evaluating a move when at least one possibility has
been found that proves the move to be worse than a previously
examined move. Such moves need not be evaluated further. When
applied to a standard minimax tree, it returns the same move as
minimax would, but prunes away branches that cannot possibly
influence the final decision. Alpha-beta search proceeds in a depth-
first fashion. An alpha value is an initial or temporary value
associated with a MAX node. Because MAX nodes are given the
maximum value among their children, an alpha value can never
decrease; it can only go up. A beta value is an initial or temporary

value associated with a MIN node. Because MIN nodes are given the minimum value among their children, a beta value can never increase; it can only go down.

For example, suppose a MAX node's alpha = 6. Then the search needn't consider any branches emanating from a MIN descendant that has a beta value that is less-than-or-equal to 6. So if you know that a MAX node has an alpha of 6, and you know that one of its MIN descendants has a beta that is less than or equal to 6, you needn't search any further below that MIN node. This is called alpha pruning.

The reason is that no matter what happens below that MIN node, it cannot take on a value that is greater than 6. So its value cannot be propagated up to its MAX (alpha) parent. Similarly, if a MIN node's beta value = 6, you needn't search any further below a descendant MAX that has acquired an alpha value of 6 or more. This is called beta pruning. The reason again is that no matter what happens below that MAX node, it cannot take on a value that is less than 6. So its value cannot be propagated up to its MIN (beta) parent.

### Rules for Alpha-beta Pruning

**Alpha Pruning**: Search can be stopped below any MIN node having a beta value less than or equal to the alpha value of any of its MAX ancestors.

**Beta Pruning**: Search can be stopped below any MAX node having a alpha value greater than or equal to the beta value of any of its MIN ancestors.

The algorithm is further illustrated by using a pseudo code:

```
function alphabeta(node, depth, α, β, Player)
    if depth = 0 or node is a terminal node
        return the heuristic value of node
    if Player = MaxPlayer
```

```
        for each child of node
            α := max(α, alphabeta(child, depth-1, α, β, not(Player) ))
            if β ≤ α
                break                    (* Beta cut-off *)
        return α
    else
        for each child of node
            β := min(β, alphabeta(child, depth-1, α, β, not(Player) ))
            if β ≤ α
                break                    (* Alpha cut-off *)
        return β
    (* Initial call *)
    alphabeta(origin, depth, -infinity, +infinity, MaxPlayer)
```

## Constraint Satisfaction Problems

Search can be made easier in cases where the solution instead of corresponding to an optimal path, is only required to satisfy local consistency conditions. We call such problems Constraint Satisfaction (CS) Problems. For example, in a crossword puzzle it is only required that words that cross each other have the same letter in the location where they cross. It would be a general search problem if we require, say, that we use at most 15 vowels. Many problems can be stated as constraints satisfaction problems. The CS approach has been used in a variety of situations, for example, in Sketchpad[Sutherland,64], an old and seminal graphical system, in Garnet[Myers,89], a recent graphical user interface, in ThingLab[Freeman-Benson,90], an object-oriented simulation system, in picture understanding [Waltz], in cryptography, temporal reasoning, in active data bases. Here are some simple examples of constraint satisfaction problems:

Example 1: The n-Queen problem: The local condition is that no two queens attack each other, i.e. are on the same row, or column, or diagonal.

Example 2: A crossword puzzle: We are to complete the puzzle by joining the available letters to form legitimate words.

Example 3: A cryptography problem

Example 4: A map coloring problem: We are given a map, i.e. a planar graph, and we are told to color it using three colors, green, red, and blue, so that no two neighboring countries have the same color.

All these examples are instances of the same pattern, captured by the following definition:

A Constraint Satisfaction Problem is characterized by:

- a set of variables $\{x_1, x_2, .., x_n\}$,

- for each variable $x_i$ a domain $D_i$ with the possible values for that variable, and

- a set of constraints, i.e. relations, that are assumed to hold between the values of the variables. [These relations can be given intentionally, i.e. as a formula, or extensionally, i.e. as a set, or procedurally, i.e. with an appropriate generating or recognizing function.] We will only consider constraints involving one or two variables.

The constraint satisfaction problem is to find, for each i from 1 to n, a value in $D_i$ for $x_i$ so that all constraints are satisfied.

A CS problem can easily be stated [Freuder] as a sentence in first order logic, of the form:

(exist $x_1$)..(exist $x_n$) ($D_1(x_1)$ & .. $D_n(x_n)$ => $C_1..C_m$)

A CS problem is usually represented as an undirected graph, called Constraint Graph where the nodes are the variables and the edges are the binary constraints. Unary constraints can be disposed of by just redefining the domains to contain only the values that satisfy all the unary constraints. Higher order constraints are represented by hyper-arcs. These problems are solved by reducing the remaining unknowns in the graph and then ultimately getting the right values for each variable stated in the problem.

# Knowledge Representation

Knowledge representation (KR) is an area of artificial intelligence research aimed at representing knowledge in symbols to facilitate inference from those knowledge elements, creating new elements of knowledge. The KR can be made to be independent of the underlying knowledge model or knowledge base system (KBS) such as a semantic network. Knowledge can be shared less ambiguously in its explicit form and this became especially important when machines started to be applied to facilitate knowledge management.

Nowadays, Knowledge Representation is a multidisciplinary field that applies theories and techniques from:

Logic: provides the formal structure and rules of inference, more details in the Logic section.

Ontology: defines the kinds of things that exist in the application domain, see the Ontology section.

Computation: supports the applications that distinguish knowledge representation from pure philosophy.

Therefore, Knowledge Representation can be defined as the application of logic and ontology to the task of constructing computable models of some domain. Logic and Ontology provide the formalization mechanisms required to make expressive models easily sharable and computer aware. Finally, thanks to computational resources, great quantities of knowledge expressed this way can be automated. Thus, the full potential of knowledge accumulations can be exploited. However, computers play only the role of powerful processors of more or less rich information sources. The final interpretation of the results is carried out by the agents that motivate these processing, in this case human users of the knowledge management systems.

At this point, it is important to remark that the possibilities of the application of actual Knowledge Representation techniques are enormous. Knowledge is always more than the sum of its parts and Knowledge Representation provides the tools needed to manage accumulations of knowledge and the World Wide Web is becoming the biggest accumulation of knowledge ever faced by humanity. These possibilities will be more deeply explored in the next State of the Art sections, devoted to Web Technologies and Semantic Web.

## Principles

In addition to the previous definition, Knowledge Representation can be also described by the five fundamental roles that it plays in artificial intelligence; they are the Knowledge Representation principles:

- A knowledge representation is a surrogate: symbols are used to represent external things that cannot be stored in a computer, i.e. physical objects, events, and relationships. Symbols are surrogates for the external things. Symbols and links between them form a model of the external system that can be manipulated to simulate it or reason about it.

- A knowledge representation is a set of ontological commitments: Ontology is the study of existence. Thus, ontology determines the categories of things that exist or may exist in an application domain. Those categories set the ontological commitments of the application designer or knowledge engineer.

- A knowledge representation is a fragmentary theory of intelligent reasoning: to support reasoning about modeled things in a domain, a knowledge representation must describe their behavior and interactions. The description constitutes a theory of the application domain. It can be stated, for instance, as explicit axioms or compiled into computable programs.

- A knowledge representation is a medium for efficient computation: besides representing knowledge, an Artificial Intelligence System must encode knowledge in a form that can be processed efficiently by the available computing equipment. Therefore, developments in computer hardware and programming theory have a great influence on knowledge representation.

- A knowledge representation is a medium for human expression: a good knowledge representation language should facilitate communication between the knowledge engineers who manage knowledge tools and the domain experts who understand the application domain. Domain experts should be able to read and verify the domain definitions and rules written by knowledge engineers.

## Logic

This is one of the fundamental aspects of knowledge representation as presented in the Knowledge Representations section. Logic was developed as an attempt to create a universal language based on mathematical principles. Therefore, it is based on formal principles that impose some requirements over a knowledge representation language to be logic:

Vocabulary it is a collection of symbols represented as chars, words, icons, or even sounds. These symbols are divided in four groups:

Logical symbols: they are domain-independent, e.g. quantifiers like "∀" or connectives like "∧".

Constants: these are domain dependent and identify individuals, properties or relations in the application domain, or universe of discourse. E.g. "truck281" or "fatherOf".

Variables: they are unbounded symbols whose range of application is governed by quantifiers.

Punctuation: these are utility symbols that separate or group other symbols, e.g. commas and parenthesis.

Syntax: a logic must have grammar rules that determine how symbols combine to form well-formed sentences.

Semantics: it is necessary to make meaningful statements. It comprises a theory of reference that determines how the constants and variables relate to things in the universe of discourse. Moreover, it also includes a theory of truth to distinguish true statements from false. More details about semantics are presented in the Semantics section.

Inference: this aspect is important in order to get something more than a notation. Inference is carried out by rules that determine how patterns are generated from others. Appropriate inference rules allow reasoning mechanisms automation and, thus, the generation of new knowledge from previous one. This point is detailed in the Inference section.

Natural Languages can represent a wider range of knowledge; however, logic enables the precisely formulated subset to be expressed in computable form. On the other hand, although there are some kinds of knowledge not expressible in logic, such knowledge cannot be represented either on any digital computer in any other notation. The expressible power of logic includes every kind of information storable or programmed on any digital computer.

# First Order Logic

Whereas propositional logic assumes world contains facts, first-order logic (like natural language) assumes the world contains

- Objects: people, houses, numbers, theories, Sachin tendulkar, colors, baseball games, wars, centuries etc.

- Relations: red, round, bogus, prime, multistoried, is the brother of, is bigger than, is inside, is part of, has color, occurred after, owns, comes between and many more

- Functions: father of, best friend, third inning of, one more than, end of and so on.

**Basic Elements of FOL:**

**Constant symbols**: Siddharth; 2; France

**Predicate symbols**: Brother; >; Likes

**Function symbols:** Sqrt; LeftLegOf ; StudentOf

**Variable symbols**:  x; y; a; b

**Connectives**:      ^ _ : ) ,

**Equality**:         =

**Quantifiers**:     $\forall$ $\exists$ also shown as A and E in below texts

**Punctuation**:    ( ) ; { }

## Atomic sentences

Atomic sentence = predicate(term1; : : : ; termn)

or term1 = term2

Term = function(term1; : : : ; termn)

or constant or variable

E.g.,

Brother(Tom Hardley; Sam Jhonson)

> (Length(LeftLegOf (Tom)); Length(LeftLegOf (Sam)))

## Complex Sentences

Sibling(Tom; Sam) ) Sibling(Sam; Tom)

>(1; 2)  ¬< (1; 2)

## Quantifiers

Universal quantification corresponds to conjunction ("and") in that (Ax)P(x) means that P holds for all values of x in the domain associated with that variable.

E.g., (Ax) dolphin(x) => mammal(x)

Existential quantification corresponds to disjunction ("or") in that (Ex)P(x) means that P holds for some value of x in the domain associated with that variable.

E.g., (Ex) mammal(x) ^ lays-eggs(x)

Universal quantifiers are usually used with "implies" to form "if-then rules."

E.g., (Ax) cs731-student(x) => smart(x) means "All cs731 students are smart."

You rarely use universal quantification to make blanket statements about every individual in the world: (Ax)cs731-student(x) ^ smart(x) meaning that everyone in the world is a cs731 student and is smart.

**English to FOL**

- Every gardener likes the sun.
  (Ax) gardener(x) => likes(x,Sun)

- You can fool some of the people all of the time.
  (Ex)(At) (person(x) ^ time(t)) => can-fool(x,t)

- You can fool all of the people some of the time.
  (Ax)(Et) (person(x) ^ time(t) => can-fool(x,t)

- All purple mushrooms are poisonous.
  (Ax) (mushroom(x) ^ purple(x)) => poisonous(x)

# Inference Rule in FOL

Inference rules for Predicate language apply to FOL as well like Modus Ponens, And-Introduction, And-Elimination, etc. New sound inference rules for use with quantifiers:

## Universal Elimination
If (Ax)P(x) is true, then P(c) is true, where c is a constant in the domain of x. For example, from (Ax)eats(Jessica, x) we can infer eats(Jessica, Ice-Cream).

The variable symbol can be replaced by any ground term, i.e., any constant symbol or function symbol applied to ground terms only.

## Existential Introduction
If P(c) is true, then (Ex)P(x) is inferred.

For example, from eats(Jessica, Ice-Cream) we can infer (Ex)eats(Jessica, x).

All instances of the given constant symbol are replaced by the new variable symbol. Note that the variable symbol cannot already exist anywhere in the expression.

**Existential Elimination**
From (Ex)P(x) infer P(c).

For example, from (Ex)eats(Jessica, x) infer eats(Jessica, Cheese).

Note that the variable is replaced by a brand new constant that does not occur in this or any other sentence in the Knowledge Base. In other words, we don't want to accidentally draw other inferences about it by introducing the constant. All we know is there must be some constant that makes this true, so we can introduce a brand new one to stand in for that (unknown) constant.

# GMP (Generalized Modus Pones)

Combines And-Introduction, Universal-Elimination, and Modus Ponens

E.g.: from P(c), Q(c), and (Ax)(P(x) ^ Q(x)) => R(x), derive R(c)

In general, given atomic sentences P1, P2, ..., PN, and implication sentence (Q1 ^ Q2 ^ ... ^ QN) => R, where Q1, ..., QN and R are atomic sentences, and subst(Theta, Pi) = subst(Theta, Qi) for i=1,...,N, derive new sentence: subst(Theta, R)

subst(Theta, alpha) denotes the result of applying a set of substitutions defined by Theta to the sentence alpha

A substitution list Theta = {v1/t1, v2/t2, ..., vn/tn} means to replace all occurrences of variable symbol vi by term ti.

Substitutions are made in left-to-right order in the list.

E.g.: subst({x/IceCream, y/Jessica}, eats(y,x)) = eats(Jessica, IceCream)

Generalized Modus Ponens (GMP) is complete for KBs containing only Horn clauses

A Horn clause is a sentence of the form:
(Ax) (P1(x) ^ P2(x) ^ ... ^ Pn(x)) => Q(x)
where there are 0 or more Pi's, and the Pi's and Q are positive (i.e., un-negated) literals

Horn clauses represent a subset of the set of sentences representable in FOL. For example, P(a) v Q(a) is a sentence in FOL but is not a Horn clause.

Natural deduction using GMP is complete for KBs containing only Horn clauses. Proofs start with the given axioms/premises in KB, deriving new sentences using GMP until the goal/query sentence is derived. This defines a forward chaining inference procedure because it moves "forward" from the KB to the goal.

## Forward Chaining

Example: KB = All cats like fish, cats eat everything they like, and Jessica is a cat. In FOL, KB =

(Ax) cat(x) => likes(x, Fish)

(Ax)(Ay) (cat(x) ^ likes(x,y)) => eats(x,y)

cat(Jessica)

Goal query: Does Jessica eat fish?.  Proof:

Use GMP with (1) and (3) to derive: 4. likes(Jessica, Fish)

Use GMP with (3), (4) and (2) to derive eats(Jessica, Fish)

So, Yes, Jessica eats fish.

## Backward Chaining

Backward-chaining deduction using GMP is complete for KBs containing only Horn clauses. Proofs start with the goal query, find implications that would allow you to prove it, and then prove each of the antecedents in the implication, continuing to work "backwards" until we get to the axioms, which we know are true.

Example: Does Jessica eat fish?

To prove eats(Jessica, Fish), first see if this is known from one of the axioms directly. Here it is not known, so see if there is a Horn clause that has the consequent (i.e., right-hand side) of the implication matching the goal.

Proof:   Goal Driven

Goal matches RHS of Horn clause (2), so try and prove new sub-goals cat(Jessica) and likes(Jessica, Fish) that correspond to the LHS of (2)

cat(Jessica) matches axiom (3), so we've "solved" that sub-goal

likes(Jessica, Fish) matches the RHS of (1), so try and prove cat(Jessica)

cat(Jessica) matches (as it did earlier) axiom (3), so we've solved this sub-goal

There are no unsolved sub-goals, so we're done. Yes, Jessica eats fish.

## Unification

Let p and q be sentences in first-order logic.

UNIFY(p,q) = U where subst(U,p) = subst(U,q)

Where subst(U,p) means the result of applying substitution U on the sentence p. Then U is called a unifier for p and q. The unification of p and q is the result of applying U to both of them.[1]

Let L be a set of sentences, for example, L = {p,q}. A unifier U is called a most general unifier for L if, for all unifiers U' of L, there exists a substitution s such that applying s to the result of applying U to L gives the same result as applying U' to L:

subst(U',L) = subst(s,subst(U,L)).

## Resolution

In first order logic, resolution condenses the traditional syllogisms of logical inference down to a single rule.

To understand how resolution works, consider the following example syllogism of term logic:

All Greeks are Europeans, Homer is a Greek.

Therefore, Homer is a European:

$$Ax.P(x) => Q(X)$$

P(a) Therefore, Q(a).

To recast the reasoning using the resolution technique, first the clauses must be converted to conjunctive normal form. In this form, all quantification becomes implicit: universal quantifiers on variables (X, Y) are simply omitted as understood, while existentially-quantified variables are replaced by Skolem functions.

$$\neg P(x) \lor Q(x)$$

P(a) Therefore, Q(a). So the question is, how does the resolution technique derive the last clause from the first two? The rule is simple:

- Find two clauses containing the same predicate, where it is negated in one clause but not in the other.

- Perform unification on the two predicates if possible.

- If any unbound variables which were bound in the unified predicates also occur in other predicates in the two clauses, replace them with their bound values (terms) there as well.

- Discard the unified predicates, and combine the remaining ones from the two clauses into a new clause, also joined by the "V" operator.

To apply this rule to the above example, we find the predicate P occurs in negated form $\neg P(X)$ in the first clause, and in non-negated form P(a) in the second clause. X is an unbound variable, while a is a bound value (term). Unifying the two produces the substitution $X \mapsto a$. Discarding the unified predicates, and applying this substitution to the remaining predicates (just Q(X), in this case), produces the conclusion: Q(a)

For another example, consider the syllogistic form

All Cretans are islanders and all islanders are liars. Therefore all Cretans are liars. Or more generally,

$\forall X \; P(X) \rightarrow Q(X), \; \forall X \; Q(X) \rightarrow R(X)$

Therefore, $\forall X \; P(X) \rightarrow R(X)$. In CNF (Conjunctive normal form), the antecedents become: $\neg P(X) \lor Q(X), \; \neg Q(Y) \lor R(Y)$

Now, unifying Q(X) in the first clause with $\neg Q(Y)$ in the second clause means that X and Y become the same variable anyway. Substituting this into the remaining clauses and combining them gives the conclusion: ¬P(X) ∨ R(X). The resolution rule, as defined by Robinson, also incorporated factoring, which unifies two literals in the same clause, before or during the application of resolution as defined above. The resulting inference rule is refutation complete, in that a set of clauses is unsatisfiable if and only if there exists a derivation of the empty clause using resolution alone.

## FOL application

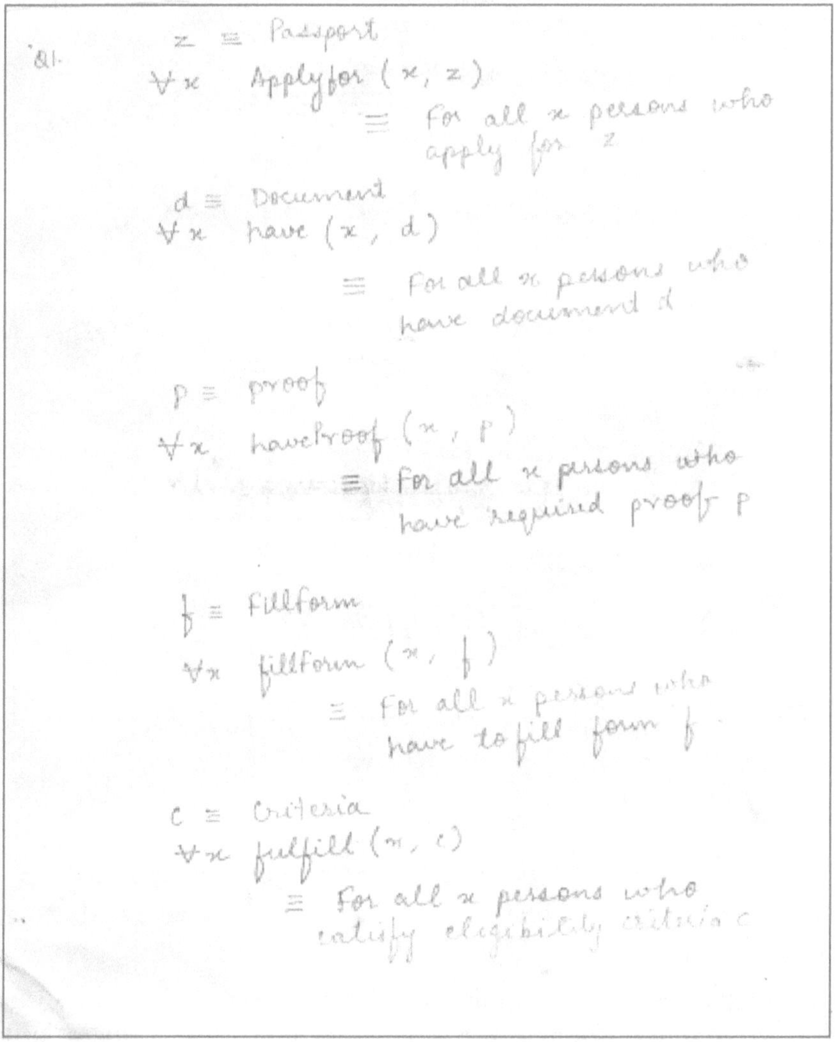

Fig.1

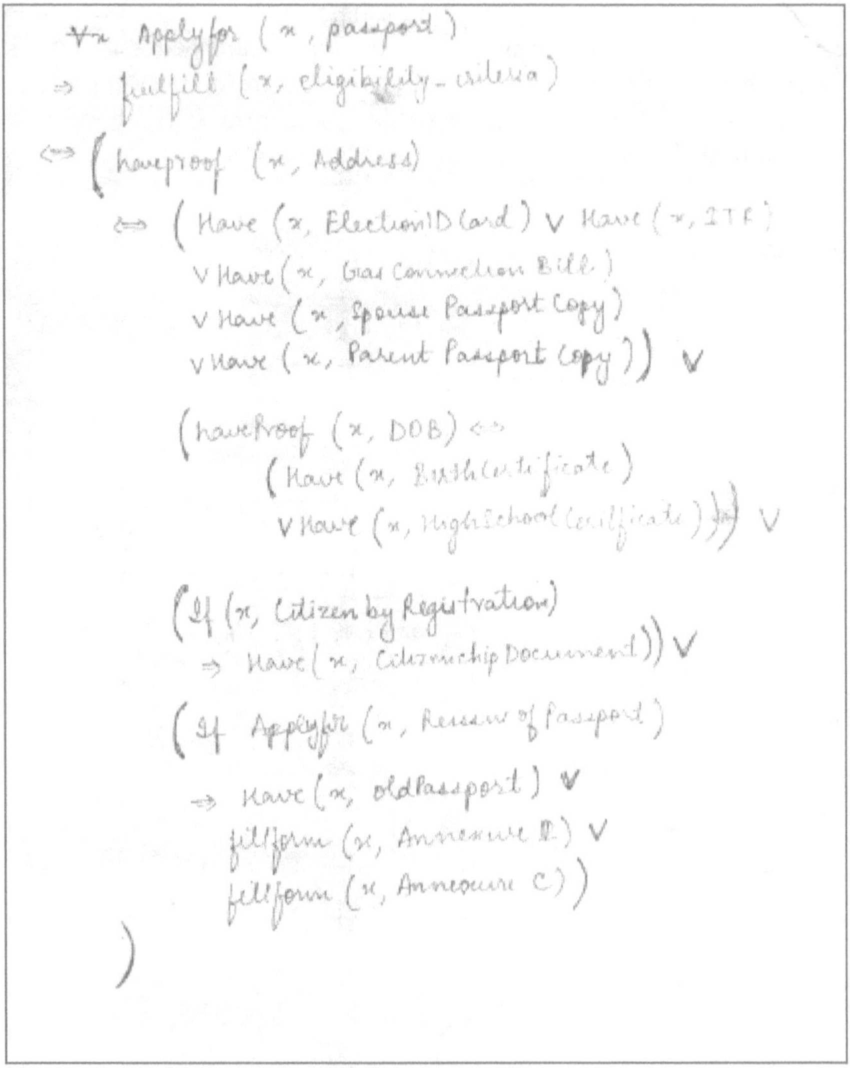

∀x Applyfor (x, passport)
⇒ fulfill (x, eligibility-criteria)
⟺ ( haveproof (x, Address)
⟺ ( Have (x, ElectionID Card) ∨ Have (x, ITF)
∨ Have (x, Gas Connection Bill)
∨ Have (x, Spouse Passport Copy)
∨ Have (x, Parent Passport Copy)) ∨

( haveProof (x, DOB) ⟺
( Have (x, Birth Certificate)
∨ Have (x, HighSchool Certificate))) ∨

( If (x, Citizen by Registration)
⇒ Have (x, Citizenship Document)) ∨

( If Applyfor (x, Renew of Passport)
⇒ Have (x, oldPassport) ∨
fillform (x, Annexure I) ∨
fillform (x, Annexure C))

)

Fig.2

The implementation of first order logic is now demonstrated using real life examples:

1) To prepare a passport application for your country, identify the rules determining eligibility for a passport and translate them into first order logic.

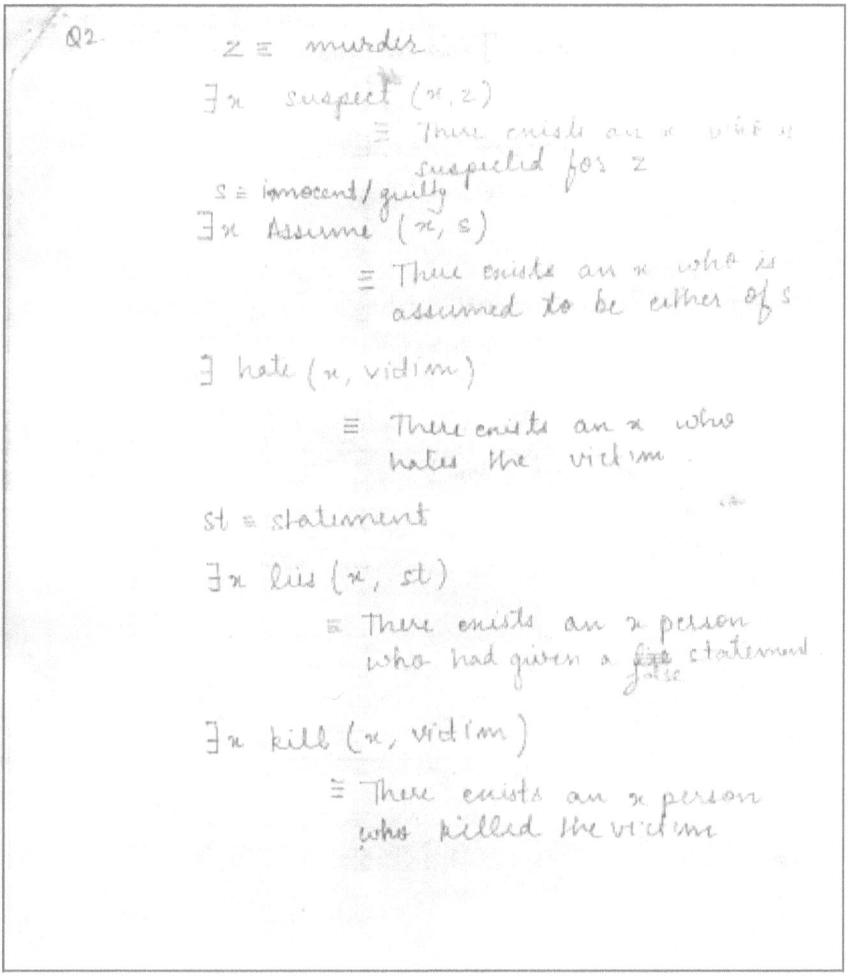

Q2.

z ≡ murder

∃x suspect (x, z)

≡ There exists an x who is suspected for z

s ≡ innocent / guilty

∃x Assume (x, s)

≡ There exists an x who is assumed to be either of s

∃ hate (x, victim)

≡ There exists an x who hates the victim.

st ≡ statement

∃x lies (x, st)

≡ There exists an x person who had given a ~~fix~~ false statement.

∃x kill (x, victim)

≡ There exists an x person who killed the victim

Fig.3

2) Write out the facts as sentences in Predicate Logic, and use Predicate resolution to solve the crime.

1. There are three suspects for a murder: Adams, Brown and Clark.

2. Adams says "I didn't do it. The victim was old acquaintance of Brown's. But Clark hated him."

3. Brown states "I didn't do it. I didn't know the guy. Besides I was out of town all the week."

4. Clark says "I didn't do it. I saw both Adams and Brown downtown with the Victim that day; one of them must have done it."

5. Assume that the two innocent men are telling the truth, but that the guilty man might be lying.

The FOL solution to the first example is shown in Fig,1 and Fig.2 where Fig.1 shows the relevant relations and Fig.2 demonstrates the application of FOL to solve the problem. Similarly, Fig.3 contains the relations required for second example and Fig.4 states out the facts which in turn the solution of the second problem.

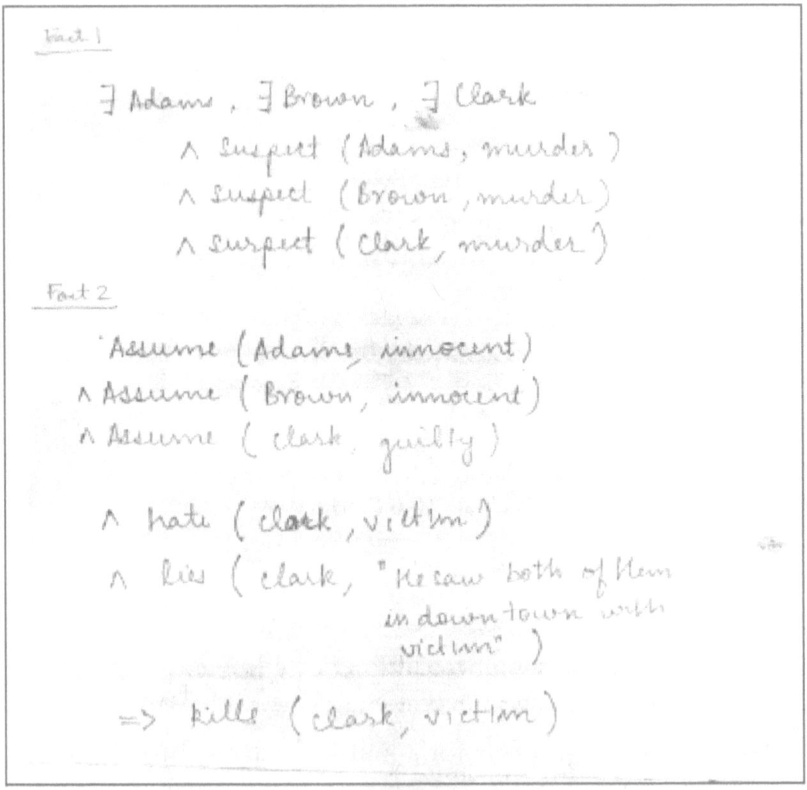

Fact 1

∃ Adams , ∃ Brown , ∃ Clark
∧ suspect (Adams, murder )
∧ suspect (Brown , murder)
∧ suspect (Clark, murder )

Fact 2

Assume (Adams, innocent )
∧ Assume (Brown, innocent )
∧ Assume (clark, guilty )

∧ hate (clark, victim )
∧ lies (clark, "He saw both of them
         in downtown with
         victim" )

⇒ kills (clark, victim )

Fig.4

These applications clearly show hoe useful first order logic is in knowledge representation and also in other fields of AI.

# Planning

## Introduction

Planning is about how an agent achieves its goals. To achieve anything but the simplest goals, an agent must reason about its future. Because an agent does not usually achieve its goals in one step, what it should do at any time depends on what it will do in the future. What it will do in the future depends on the state it is in, which, in turn, depends on what it has done in the past. Classical planning has the following assumptions:

- The agent's actions are deterministic; that is, the agent can predict the consequences of its actions.

- There are no external events beyond the control of the agent that change the state of the world.

- The world is fully observable; thus, the agent can observe the current state of the world.

- Time progresses discretely from one state to the next.

- Goals are predicates of states that must be achieved or maintained.

## Representing States, Actions and Goals

To reason about what to do, an agent must have goals, some model of the world, and a model of the consequences of its actions. A deterministic action is a partial function from states to states. It is partial because not every action can be carried out in every state. The precondition of an action specifies when the action can be carried out. The effect of an action specifies the resulting state.

### Explicit State-Space Representation

One possible representation of the effect and precondition of actions is to explicitly enumerate the states and, for each state, specify the actions that are possible in that state and, for each state-action pair, specify the state that results from carrying out the action in that state. The resulting representation is a finite state automaton.

This is not a practical approach for a domain with a large number of states and a large number of possible actions at each state.

## Feature-Based Representation of Actions

A feature-based representation of actions models

- which actions are possible in a state, in terms of the values of the features of the state,

- how the feature values in the next state are affected by the feature values of the current state and the action.

For example, in the block problem, each block is characterized by two features on(x,y) and clear(x). The feature-based representation of actions uses rules to specify the value of each variable in the state resulting from an action. The bodies of these rules can include the action carried out and the values of features in the previous state.

The rules have two forms:

- A causal rule specifies when a feature gets a new value.

- A frame rule specifies when a feature keeps its value.

It is useful to think of these as two separate cases: what makes the feature change its value, and what makes it keep its value. In the block problem, the causal rules are represented as post-effects of the actions, while the frame rules are implicit: a feature does not change its value unless a causal rule applies that changes its value. Causal rules and frame rules do not specify when an action is possible. What is possible is defined by the precondition of the actions. The precondition of an action is a proposition that must be true before the action can be carried out. In terms of constraints, the robot is constrained to only be able to choose an action for which the precondition is true.

For example, in the block problem move(x,y) has preconditions clear(x) and clear(y). The action move(x,table) is always possible. When the feature-based representation uses propositions, e.g. on(a, table), it is sometimes useful to model the effects by ADD and DELETE meta-actions. DELETE removes those propositions that are made false by an action, and ADD adds ne propositions that are made true by an action.

Example:

Previous state: on(a,b), clear(b), clear(c)3

Action: move(a,c)

Meta-actions: delete(clear(c)), delete((on(a,b)), add(on(a,c)), add(clear(b)).

A deterministic plan is a sequence of actions to achieve a goal from a given starting state. A deterministic planner is a problem solver that can produce a plan. The input to a planner is an initial world description, a specification of the actions available to the agent, and a goal description. The specification of the actions includes their preconditions and their effects. One of the simplest planning strategies is to treat the planning problem as a path planning problem in the state-space graph. In a state-space graph, nodes are states, and arcs correspond to actions from one state to another. The arcs coming out of a state **s** correspond to all of the legal actions that can be carried out in that state. That is, for each state **s**, there is an arc for each action a whose precondition holds in state **s**, and where the resulting state does not violate a maintenance goal which is a proposition that must be true in every state through which the agent passes. These are often safety goals - the goal of staying away from bad states. A plan is a path from the initial state to a state that satisfies the achievement goal that is a proposition which must be true in the final state of the planning algorithm.

It is also possible to search backward from the set of states that satisfy the goal. This approach is not practical when a large number of states satisfy the goal. Example: in the block problem a very large number of states satisfy the goal **on(a,b)**. It is often more efficient to search in a different search space - one where the nodes are not states but rather are goals to be achieved.

Regression planning is searching in the graph defined by the following:

- The nodes are goals that must be achieved. A goal is a set of assignments to some of the features.

- The arcs correspond to actions. In particular, an arc from node g to g', labeled with action act, means act is the last action that is carried out before goal g is achieved, and the node g' is the goal that must be true immediately before act so that g is true  immediately after act.

- The start node is the goal to be achieved. Here we assume it is a conjunction of assignments of values to features.

- The goal condition for the search, goal(g), is true if all of the elements of g are true of the initial state.

## Partial Order Planning

The forward and regression planners enforce a total ordering on actions at all stages of the planning process. The idea of a partial-order planner is to have a partial ordering between actions and only commit to an ordering between actions when forced. A partial ordering is a less-than relation that is transitive and asymmetric. A partial-order plan is a set of actions together with a partial ordering, representing a "before" relation on actions, such that any total ordering of the actions, consistent with the partial ordering, will solve the goal from the initial state. Write act0 < act1 if action act0 is before action act1 in the partial order. This means that action act0 must occur before action act1. For uniformity, we use two pseudo actions start and finish.

- Start is an action that achieves the relations that are true in the initial state, and finish is an action whose precondition is the goal to be solved.

- The pseudo action start is before every other action, and finish is after every other action.

The use of these as actions means that the algorithm does not require special cases for the initial situation and for the goals. When the preconditions of finish hold, the goal is solved. We must ensure that the actions achieve the conditions they were assigned to achieve.

- Each precondition P of an action act1 in a plan will have an action act0 associated with it such that act0 achieves precondition P for act1.

- The triple ⟨act0,P,act1⟩ is a causal link. The partial order specifies that action act0 occurs before action act1, which is written as act0 < act1.

- Any other action A that makes P false must either be before act0 or after act1.

Informally, a partial-order planner works as follows: Begin with the actions start and finish and the partial order start < finish. The planner maintains an agenda that is a set of ⟨P,A⟩ pairs, where A is an action in the plan and P is an atom that is a precondition of A that must be achieved. The execution of the partial order planning is explained in detail in the following algorithm sketch:

function POP(initial, goal, actions) returns plan
    plan←Make-Minimal-Plan(initial, goal)
    loop do
        if Solution?( plan) then return plan
        $S_{need}$, c←Select-Subgoal( plan)
        Choose-Action( plan, actions,$S_{need}$, c)
        Resolve-Threats( plan)
    end


function Select-Subgoal( plan) returns $S_{need}$, c
    pick a plan step $S_{need}$ from Steps( plan)
        with a precondition c that has not been achieved
    return $S_{need}$, c


procedure Choose-Action(plan, actions, $S_{need}$, c)
    choose a step $S_{add}$ from actions or Steps( plan) that has c as an effect
    if there is no such step then fail

add the causal link $S_{add}$ -$^c$--> Sneed to Links( plan)

add the ordering constraint $S_{add}$ < Sneed to Orderings( plan)

if $S_{add}$ is a newly added step from actions then

    add $S_{add}$ to Steps( plan)

    add Start < $S_{add}$ < Finish to Orderings( plan)

procedure Resolve-Threats(plan)

    for each $S_{threat}$ that threatens a link Si -$^c$--> Sj in Links( plan) do

        choose either

            Demotion: Add $S_{threat}$ < Si to Orderings( plan)

            Promotion: Add Sj < $S_{threat}$ to Orderings( plan)

        if not Consistent( plan) then fail

    end

Algorithm can be understood efficiently by observing the following example which is demonstrated in three parts and arranged in a diagrammatic representation with nodes arrows indicating the flow of the plan by completing the actions required to achieve the desired goal:

**Initial State:**

**Middle State:**

This stage is the result of implementing few actions and then observing their output and the level of completeness of the plan.

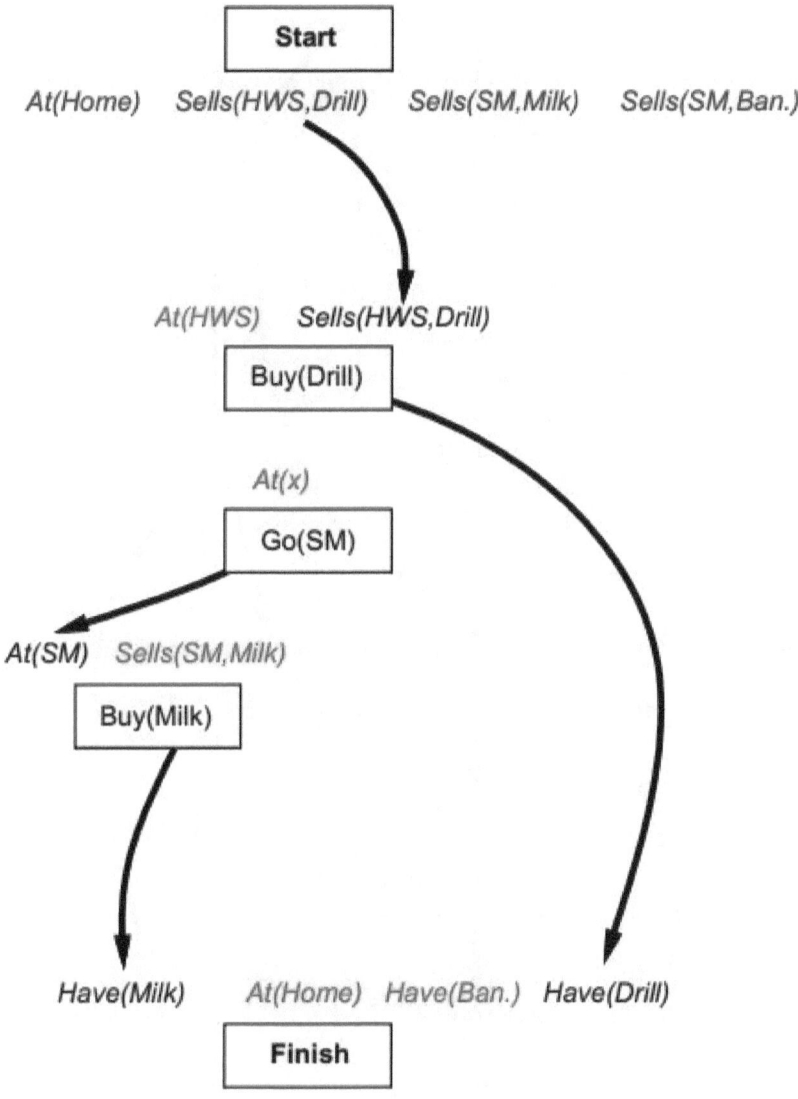

**Final State:**

This is the final representation of the plan showing the actions as nodes and their results are connected via arrows ultimately reaching the required goal state.

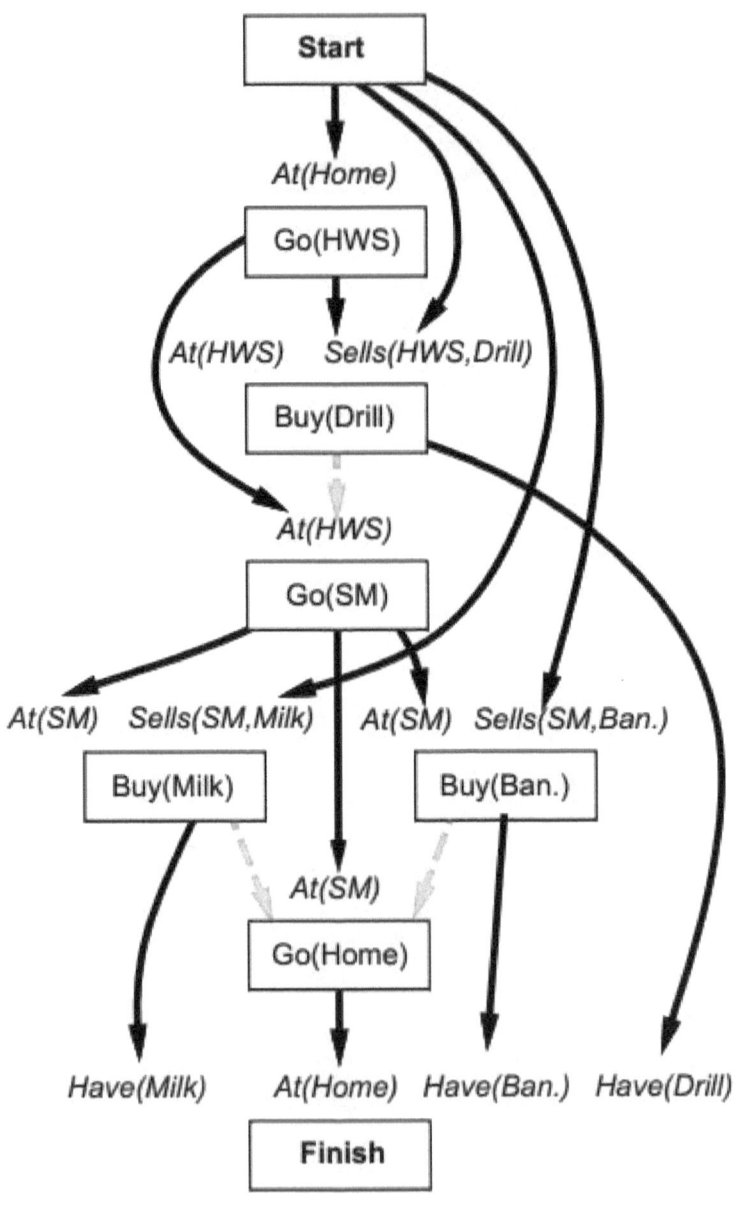

# Strips

The "classical" approach most planners use today is derived from the STRIPS language. STRIPS was devised in the early 1970s to control a robot called Shakey. The STRIPS language is very simple but expressive language that lends itself to efficient planning algorithms. Planning can be considered as a logical inference problem: a plan is inferred from facts and logical relationships. These are represented planning problems as a series of state descriptions and operators expressed in first-order predicate logic.

State descriptions represent the state of the world at three points during the plan:

- Initial state, the state of the world at the start of the problem;

- Current state, and

- Goal state, the state of the world we want to get to.

Operators are actions that can be applied to change the state of the world.

- Each operator has outcomes i.e. how it affects the world.

- Each operator can only be applied in certain circumstances. These are the preconditions of the operator.

**Example Problem**: Construct a script for going to a movie from the viewpoint of the moviegoer.

### Solution using Strips:

| Agents or Variables | |
|---|---|
| | • Show List |
| | • Movie |
| Props: | • Ticket |
| • Home | • Seat |
| • Box-office | • Seat List |
| • Money | • Entrance Door |
| • Conveyance | • Exit Door |
| • Multiplex | • Ticket Counter |
| • Movie Hall | |

Roles:
- D = Driver
- C = Conductor
- K = Cashier
- A = Attendant
- S = Owner of Multiplex
- M = I (Myself)

Entry Conditions:
- M has money
- M wants to see the movie
- Conveyance is available

Results:
- M has less money
- M have seen the movie
- M is pleased (Optional)
- S has more money
- Increment in the Box Office collection of the movie

Different scenes of the given situation are demonstrated:

**Scene 1**: *Finding conveyance to movie hall*

M CONC Conveyance
M MBUILD Conveyance
M PTRANS M to Conveyance
M MOVE M to sitting position in Conveyance
D PTRANS Conveyance from Home to Multiplex
M MOVE M to get down from Conveyance
M ATRANS Money to Conductor
M PTRANS M to the entrance of Multiplex
M PTRANS M to the ticket counter

**Scene 2:** *Buying Ticket*

M CONC Show List
*M ATTENDS Show List
M MBUILD Movie from Show List
M CONC Seat List
M ATTENDS Seat List
M MBUILD Seat from Seat List
(If M finds no Movie then go to **)
M MTRANS K for Ticket of the Movie
K SPEAK M for Money
M ATRANS Money to K
K GRASP Money
K ATRANS Ticket to M
M GRASP Ticket
M PTRANS M to Movie Hall

**Scene 3:** *Watching the Movie*
M PROPEL Entrance Door
M PTRANS M into Movie Hall
A SPEAK for Ticket
M ATRANS Ticket to A
A GRASP Ticket
A MTRANS M direction to Seat
A ATRANS Ticket to M
M GRASP Ticket
M PTRANS M to Seat
M MOVE M to sit on Seat
M ATTEND Movie
(Go back to *) or (Go to scene 4)

**Scene 4:** *Exiting*
M MOVE M to get up from Seat
M PTRANS M to Exit Door
M PROPEL Exit Door to get out
**M PTRANS M out of Movie Hall
M PTRANS M out of the MULTIPLEX
M CONC Conveyance
M MBUILD Conveyance
M PTRANS M to Conveyance
D PTRANS Conveyance from Multiplex to Home

# Semantic networks

A semantic network or net is a graphic notation for representing knowledge in patterns of interconnected nodes and arcs. Computer implementations of semantic networks were first developed for artificial intelligence and machine translation, but earlier versions have long been used in philosophy, psychology, and linguistics.

What is common to all semantic networks is a declarative graphic representation that can be used either to represent knowledge or to support automated systems for reasoning about knowledge. Some versions are highly informal, but other versions are formally defined systems of logic. Following are six of the most common kinds of semantic networks:

- Definitional networks emphasize the subtype or are-a relation between a concept type and a newly defined subtype. The resulting network, also called a generalization or subsumption hierarchy, supports the rule of inheritance for copying properties defined for a super type to all of its subtypes. Since definitions are true by definition, the information in these networks is often assumed to be necessarily true.

- Assertional networks are designed to assert propositions. Unlike definitional networks, the information in an assertional network is assumed to be contingently true, unless it is explicitly marked with a modal operator. Some assertional networks have been proposed as models of the conceptual structures underlying natural language semantics.

Implicational networks use implication as the primary relation for connecting nodes. They may be used to represent patterns of beliefs, causality, or inferences.

Executable networks include some mechanism, such as marker passing or attached procedures, which can perform inferences, pass messages, or search for patterns and associations.

Learning networks build or extend their representations by acquiring knowledge from examples. The new knowledge may change the old network by adding and deleting nodes and arcs or by modifying numerical values, called weights, associated with the nodes and arcs. Hybrid networks combine two or more of the previous techniques, either in a single network or in separate, but closely interacting networks.

Some of the networks have been explicitly designed to implement hypotheses about human cognitive mechanisms, while others have been designed primarily for computer efficiency. Sometimes, computational reasons may lead to the same conclusions as psychological evidence. The distinction between definitional and assertional networks, for example, has a close parallel to Tulving's (1972) distinction between semantic memory and episodic memory.

Network notations and linear notations are both capable of expressing equivalent information, but certain representational mechanisms are better suited to one form or the other. Since the boundary lines are vague, it is impossible to give necessary and sufficient conditions that include all semantic networks while excluding other systems that are not usually called semantic networks.

Semantic networks are knowledge representation schemes involving nodes and links (arcs or arrows) between nodes. The nodes represent objects or concepts and the links represent relations between nodes. The links are directed and labeled; thus, a semantic network is a directed graph. In print, the nodes are usually represented by circles or boxes and the links are drawn as arrows between the circles. This represents the simplest form of a semantic network, a collection of undifferentiated objects and arrows. The structure of the network defines its meaning. The meanings are merely which node has a pointer to which other node. The network defines a set of binary relations on a set of nodes.

The use of semantics networks to represent a particular problem of representing a complex market distribution network is demonstrated in Fig.5. In this network, class nodes or generic items are shown in rectangular boxes; instances of class nodes are represented in ellipses and text strings on the arrows represent the relation between different nodes.

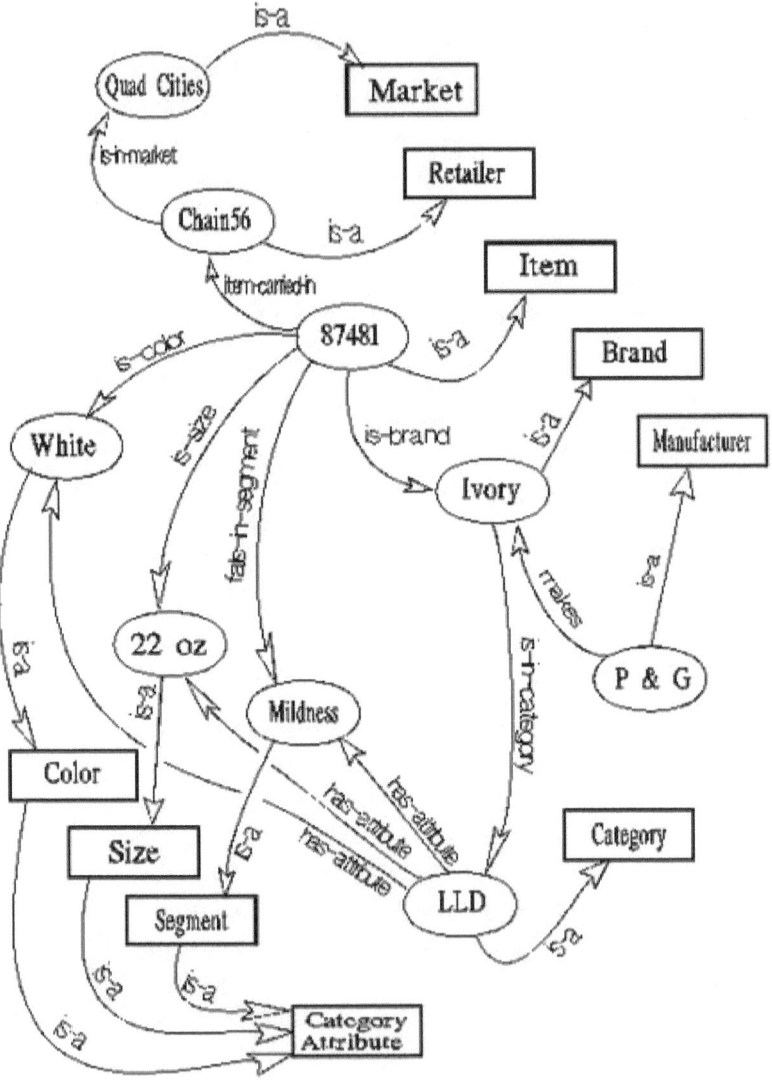

Fig.5

## Frames

Relation to semantic networks:

It becomes necessary to assign more structure to nodes, as well as to links. For example, in many cases we need node labels that can be computed, rather than being fixed in advance. It is natural to use database ideas to keep track of everything and the nodes and their relations begin to look more like frames.

A *frame* consists of a selection of slots which can be filled by values, or procedures for calculating values, or pointers to other frames. For example:

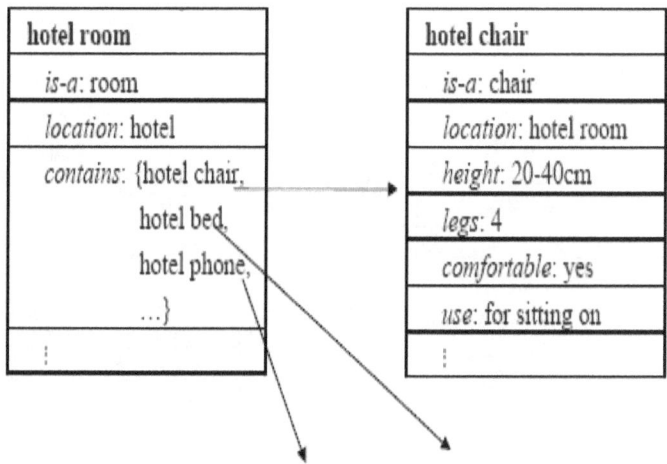

A complete frame based representation will consist of a whole hierarchy or network of frames connected together by appropriate links/pointers.

## Sample Problem:

We want to build a program that helps schedule rooms for classes of various sizes at a university, using the sort of frame technology (frames, slots, and attached procedures. Slots of frames might be used to record when and where a class is to be held, the capacity of a room, and so on, and IF-ADDED and other procedures might be used to encode constraints as well as to fill in implied values when the KB is updated.

Consider updating the KB in several ways:

a. asserting that a class of a given size is to be held in a given room at a given time; the system would either go ahead and add this to its schedule or alert the user that it was not possible to do so;

b. asserting that a class of a given size is to be held at a given time, with the system providing a suitable room (if one is available) when queried;

c. asserting that a class of a given size is desired, with the system providing a time and place when queried.

For a classroom scheduler as sketched above, below is the solution to the above questions in the form of frames:

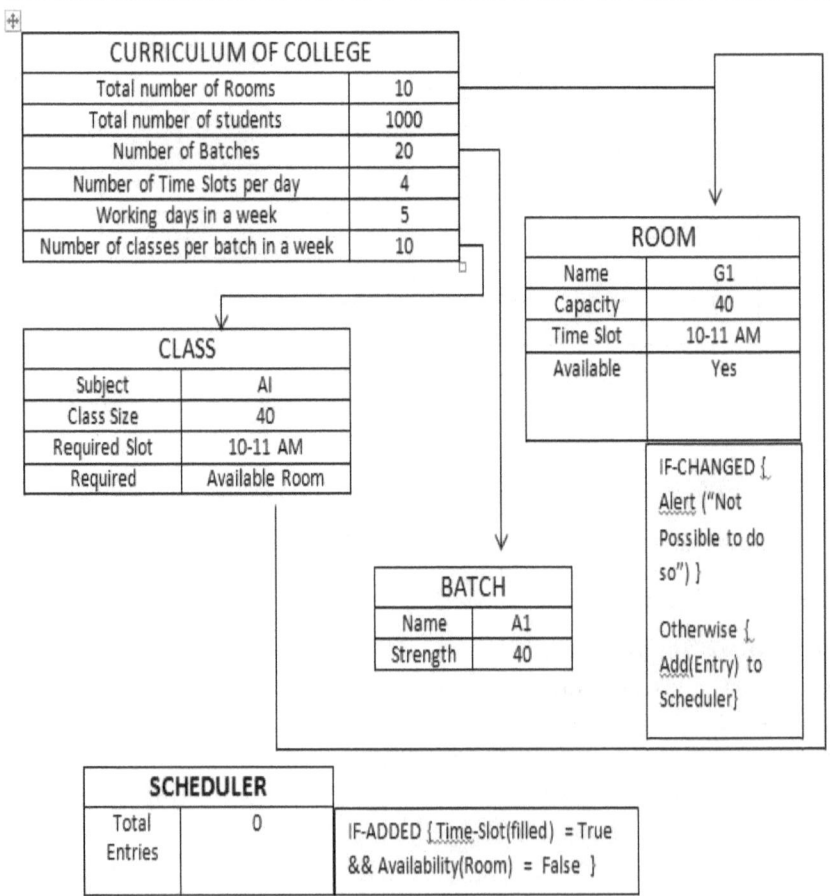

| CURRICULUM OF COLLEGE | |
| --- | --- |
| Total number of Rooms | 10 |
| Total number of students | 1000 |
| Number of Batches | 20 |
| Number of Time Slots per day | 4 |
| Working days in a week | 5 |
| Number of classes per batch in a week | 10 |

| ROOM | |
| --- | --- |
| Name | G1 |
| Capacity | 40 |
| Time Slot | 10-11 AM |
| Available | Yes |

| CLASS | |
| --- | --- |
| Subject | AI |
| Class Size | 40 |
| Required Slot | 10-11 AM |
| Required | Available Room |

IF-CHANGED {
Alert ("Not
Possible to do
so") }

Otherwise {
Add(Entry) to
Scheduler}

| BATCH | |
| --- | --- |
| Name | A1 |
| Strength | 40 |

| SCHEDULER | | |
| --- | --- | --- |
| Total Entries | 0 | IF-ADDED {Time-Slot(filled) = True && Availability(Room) = False } |

# Probabilistic reasoning

Its aim is to combine the capacity of probability theory to handle uncertainty with the capacity of deductive logic to exploit structure. The result is a richer and more expressive formalism with a broad range of possible application areas. Probabilistic logics attempt to find a natural extension of traditional logic truth tables: the results they define are derived through probabilistic expressions instead. A difficulty with probabilistic logics is that they tend to multiply the computational complexities of their probabilistic and logical components. Other difficulties include the possibility of counter-intuitive results, such as those of Dempster-Shafer theory. The need to deal with a broad variety of contexts and issues has led to many different proposals. One such is Bayesian Networks.

## Bayesian Network

Bayesian networks (BNs), also known as belief networks (or Bayes nets for short), belong to the family of probabilistic graphical models (GMs). These graphical structures are used to represent knowledge about an uncertain domain. In particular, each node in the graph represents a random variable, while the edges between the nodes represent probabilistic dependencies among the corresponding random variables. These conditional dependencies in the graph are often estimated by using known statistical and computational methods. Hence, BNs combine principles from graph theory, probability theory, computer science, and statistics.

Probabilistic graphical models are graphs in which nodes represent random variables, and the (lack of) arcs represent conditional independence assumptions. Hence they provide a compact representation of joint probability distributions. Undirected graphical models, also called Markov Random Fields (MRFs) or Markov networks, have a simple definition of independence: two (sets of) nodes A and B are conditionally independent given a third set, C, if all paths between the nodes in A and B are separated by a node in C. By contrast, directed graphical models also known as

BNs that have a more complicated notion of independence, which takes into account the directionality of the arcs. Undirected graphical models are more popular with the physics and vision communities, and directed models are more popular with the AI and statistics communities. (It is possible to have a model with both directed and undirected arcs, which is called a chain graph.) For a careful study of the relationship between directed and undirected graphical models, see the books by Pearl88, Whittaker90, and Lauritzen96.

In addition to the graph structure, it is necessary to specify the parameters of the model. For a directed model, we must specify the Conditional Probability Distribution (CPD) at each node. If the variables are discrete, this can be represented as a table (CPT), which lists the probability that the child node takes on each of its different values for each combination of values of its parents. Consider the following example, in which all nodes are binary, i.e., have two possible values, denoted by T (true) and F (false).

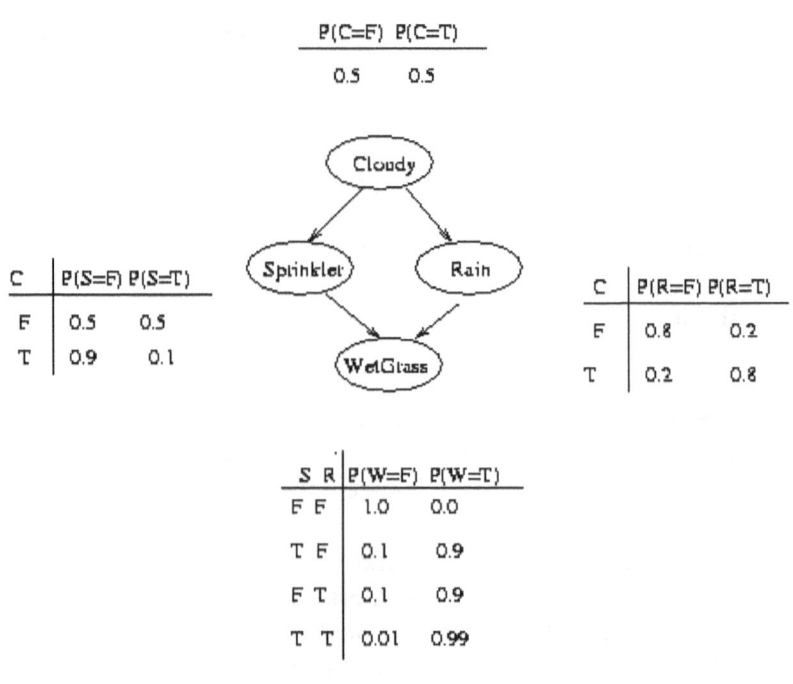

Fig. 6

In the example representation of BN shown in Fig.6, we see that the event "grass is wet" (W=true) has two possible causes: either the water sprinkler is on (S=true) or it is raining (R=true). The strength of this relationship is shown in the table. For example, we see that Pr(W=true | S=true, R=false) = 0.9 (second row), and hence, Pr(W=false | S=true, R=false) = 1 - 0.9 = 0.1, since each row must sum to one. Since the C node has no parents, its CPT specifies the prior probability that it is cloudy (in this case, 0.5). (Think of C as representing the season: if it is a cloudy season, it is less likely that the sprinkler is on and more likely that the rain is on.) The simplest conditional independence relationship encoded in a Bayesian network can be stated as follows: a node is independent of its ancestors given its parents, where the ancestor/parent relationship is with respect to some fixed topological ordering of the nodes.

By the chain rule of probability, the joint probability of all the nodes in the graph above is

$$P(C, S, R, W) = P(C) * P(S|C) * P(R|C,S) * P(W|C,S,R)$$

By using conditional independence relationships, we can rewrite this as $P(C, S, R, W) = P(C) * P(S|C) * P(R|C) * P(W|S,R)$

where we were allowed to simplify the third term because R is independent of S given its parent C, and the last term because W is independent of C given its parents S and R. We can see that the conditional independence relationships allow us to represent the joint more compactly. Here the savings are minimal, but in general, if we had n binary nodes, the full joint would require $O(2^n)$ space to represent, but the factored form would require $O(n\ 2^k)$ space to represent, where k is the maximum fan-in of a node. And fewer parameters minimize the effort of learning the BN systems. The implementation of BN used to solve probabilistic problems is further explained by the C# code shown below:

```
namespace BayesNetworkExample
{
    using System;
```

```
using BayesServer.Inference.RelevanceTree;
public static class NetworkExample
{
    public static void Main()
    {
        // create a simple Bayesian network
        var network = new Network("Demo");
        var defaultStates = new string[] {
            "True", "False" };
var a = new Node("A", defaultStates);
var b = new Node("B", defaultStates);
var c = new Node("C", defaultStates);
var d = new Node("D", defaultStates);
network.Nodes.Add(a);
network.Nodes.Add(b);
network.Nodes.Add(c);
network.Nodes.Add(d);
network.Links.Add(new Link(a, b));
network.Links.Add(new Link(a, c));
network.Links.Add(new Link(b, d));
network.Links.Add(new Link(c, d));
// Probability distribution
        var tableA = a.NewDistribution().Table;
        tableA[0] = 0.1;
         tableA[1] = 0.9;
         a.Distribution = tableA;
        var tableB = b.NewDistribution().Table;
        var iteratorB = new TableIterator(tableB,
new Node[] { a, b });
    var values = new double[] { 0.2, 0.8, 0.15,
0.85 };
   for (int i = 0; i < tableB.Count; i++,
iteratorB++)
    {
    iteratorB.Value = values[i];
            }
```

```
    b.Distribution = tableB;
    var tableC = c.NewDistribution().Table;
// we could also update the values using a
TableAccessor ( which allows random access ).
    var accessor = new TableAccessor(tableC, new
Node[] { a, c });
    accessor[0] = 0.3;
    accessor[1] = 0.7;
    accessor[2] = 0.4;
    accessor[3] = 0.6;
    c.Distribution = tableC;
    var tableD = d.NewDistribution().Table;
    var iteratorD = new TableIterator(tableD, new
Node[] { b, c, d });
    iteratorD.CopyFrom(new double[] { 0.4, 0.6,
0.55, 0.45, 0.32, 0.68, 0.01, 0.99 });
    d.Distribution = tableD;
// Save the netwotk

var factory = new RelevanceTreeInferenceFactory();
var inference =
factory.CreateInferenceEngine(network);
var queryOptions = factory.CreateQueryOptions();
var queryOutput = factory.CreateQueryOutput();
// set P(A|D=True)
inference.Evidence.SetState(d, 0);
 var queryA = new Table(a);
 inference.QueryDistributions.Add(queryA);
  inference.Query(queryOptions, queryOutput);
   Console.WriteLine("P(A|D=True) = {" + queryA[0]
+ "," + queryA[1] + "}.");
     inference.Evidence.SetState(c, 0);
     queryOptions.LogLikelihood = true;
   inference.Query(queryOptions, queryOutput);

Console.WriteLine(string.Format("P(A|D=True,
C=True) = {{{0},{1}}}, log-likelihood = {2}.",
queryA[0], queryA[1],
queryOutput.LogLikelihood.Value));
```

```
/* Expected output ...
/ P(A|D=True, C=True) =
{0.0777777777777778,0.922222222222222}, log-
likelihood = -2.04330249506396.
*/
   }  }
}
```

This code demonstrates that how a Bayesian network can be generated and then used to calculate the required probabilities from the given probability distribution.

# Dynamic Bayesian Networks

Fig.7 shows the connection of DBN (Dynamic Bayesian Networks) with BN and other models.

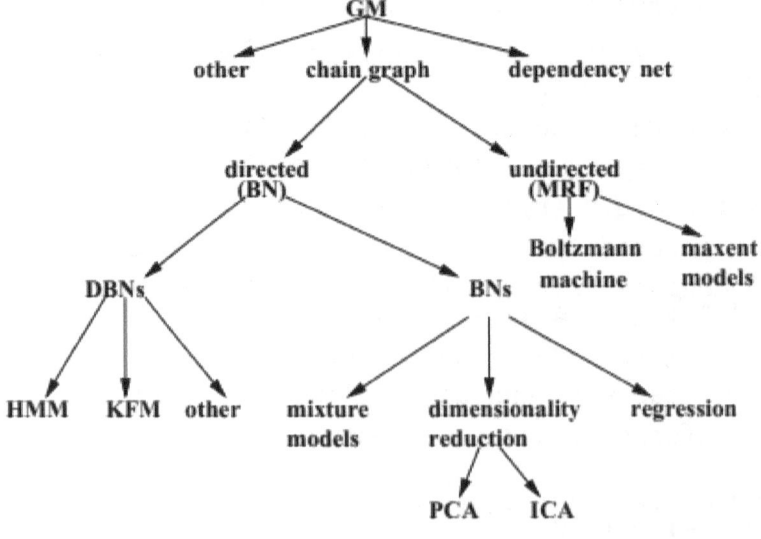

Fig. 7

## State-space models

- Assume there is some underlying hidden state of the world(query) that generates the observations(evidence), and that this hidden state evolves in time, possibly as a function of our inputs

- The belief state: our belief on the hidden state of the world given the observations up to the current time y1:t and our inputs u1:t to the system, $P( X \mid y1:t, u1:t )$

- Two most common state-space models: Hidden Markov Models(HMMs) and Kalman Filter Models(KFMs)

- a more general state-space model: dynamic Bayesian networks(DBNs)

Any state-space model must define a prior $P(X1)$ and a state-transition function, $P(Xt \mid Xt\text{-}1)$ , and an observation function, $P(Yt \mid Xt)$.

Assumptions:

- Models are first-order Markov, i.e., $P(Xt \mid X1:t\text{-}1) = P(Xt \mid Xt\text{-}1)$

- observations are conditional first-order Markov $P(Yt \mid Xt , Yt\text{-}1) = P(Yt \mid Xt)$

- Time-invariant or homogeneous

Representations:

- HMMs: Xt is a discrete random variables

- KFMs: Xt is a vector of continuous random variables

- DBNs: more general and expressive language for representing state-space models

A state-space model defines how Xt generates Yt and Xt.

The goal of inference is to infer the hidden states(query) X1:t given the observations(evidence) Y1:t.

Inference tasks:

- Filtering(monitoring): recursively estimate the belief state using Bayes' rule

  predict: computing $P(X_t \mid y_{1:t-1})$

  updating: computing $P(X_t \mid y_{1:t})$

  throw away the old belief state once we have computed the prediction("rollup")

- Smoothing: estimate the state of the past, given all the evidence up to the current time

  Fixed-lag smoothing(hindsight): computing $P(X_{t-l} \mid y_{1:t})$ where $l > 0$ is the lag

- Prediction: predict the future

  Lookahead: computing $P(X_{t+h} \mid y_{1:t})$ where $h > 0$ is how far we want to look ahead

- Viterbi decoding: compute the most likely sequence of hidden states given the data

  MPE(abduction): $x^*_{1:t} = \operatorname{argmax} P(x_{1:t} \mid y_{1:t})$

Parameters learning (system identification) in space models means estimating from data these parameters that are used to define the transition model $P(X_t \mid X_{t-1})$ and the observation model $P(Y_t \mid X_t)$. The usual criterion is maximum-likelihood after the implementation of state space models.

## Hidden Markov Models

The Hidden Markov Model (HMM) is a powerful statistical tool for modeling generative sequences that can be characterized by an underlying process generating an observable sequence. HMMs have found application in many areas interested in signal processing, and in particular speech processing, but have also been applied with success to low level NLP tasks such as part-of-speech tagging, phrase chunking, and extracting target information from documents. Andrei Markov gave his name to the mathematical theory of Markov processes in the early twentieth century, but it was Baum and his colleagues that developed the theory of HMMs in the 1960s.

**Markov Processes**: Fig.8 depicts an example of a Markov process. The mode presented describes a simple model for a stock market index. The model has three states, Bull, Bear and Even, and three index observations up, down, unchanged. The model is a finite state automaton, with probabilistic transitions between states. Given a sequence of observations, example: up-down-down we can easily verify that the state sequence that produced those observations was: Bull-Bear-Bear, and the probability of the sequence is simply the product of the transitions, in this case $0.2 \times 0.3 \times 0.3$.

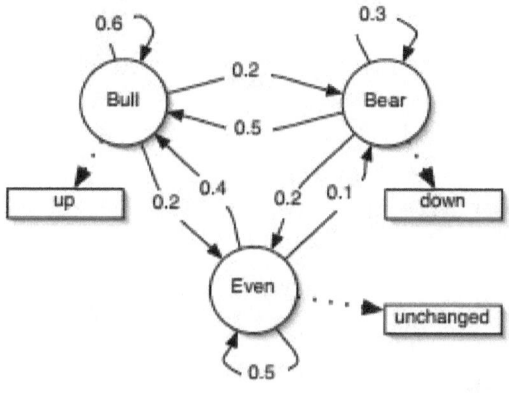

Fig.8

**Hidden Markov Models** Fig.9 shows an example of how the previous model can be extended into a HMM. The new model now allows all observation symbols to be emitted from each state with a finite probability. This change makes the model much more expressive and able to better represent our intuition, in this case, that a bull market would have both good days and bad days, but there would be more good ones. The key difference is that now if we have the observation sequence up-down-down then we cannot say exactly what state sequence produced these observations and thus the state sequence is 'hidden'. We can however calculate the probability that the model produced the sequence, as well as which state sequence was most likely to have produced the observations.

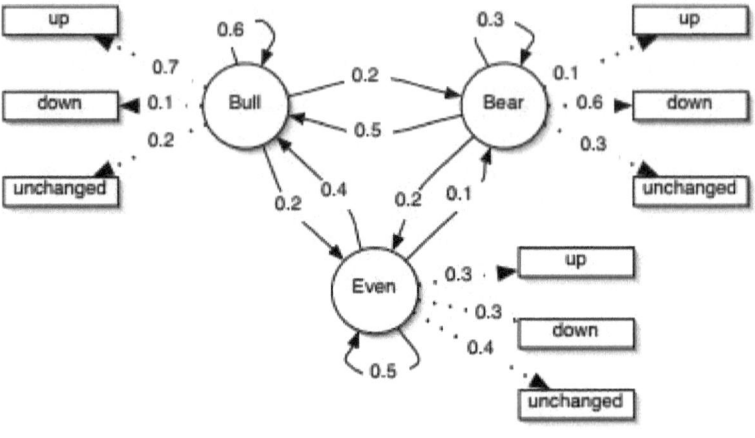

Fig. 9

## Kalman Filter Models

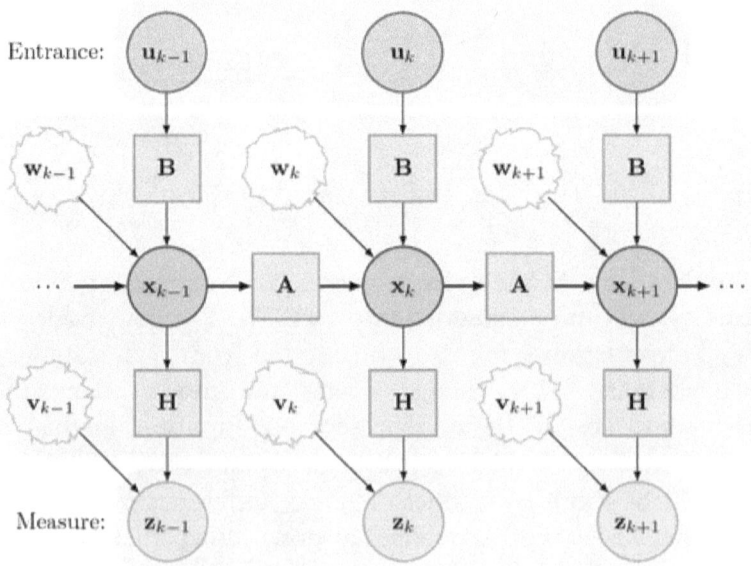

Fig.10

KFL has the same topology as an HMM, all the nodes are assumed to have linear-Gaussian distributions

$$x(t+1) = F*x(t) + w(t),$$

- $w \sim N(0, Q)$ : process noise, $x(0) \sim N(X(0), V(0))$

$$y(t) = H*x(t) + v(t),$$

- $v \sim N(0, R)$ : measurement noise

Also known as Linear Dynamic Systems(LDSs), a partially observed stochastic process with linear dynamics and linear observations: f( a + b) = f(a) + f(b) both subject to Gaussian noise. Fig.10 shows an example of Kalman Filter system model. KFL is the simplest continuous DBN and continuous state variable with linear-Gaussian dynamics and measurements.

## Features of DBN

- DBNs are directed graphical models of stochastic process
- DBNs generalize HMMs and KFLs by representing the hidden and observed state in terms of state variables, which can have complex interdependencies
- The graphical structure provides an easy way to specify these conditional independencies
- A compact parameterization of the state-space model
- An extension of BNs to handle temporal models
- Time-invariant: the term 'dynamic' means that we are modeling a dynamic model, not that the networks change over time

A DBN is defined as a pair (B0, B1:n), where B0 defines the prior P(Z1), and is a two-slice temporal Bayes net(2TBN) which defines P(Zt | Zt-1) by means of a DAG(directed acyclic graph) as follows:

$$P(Z_t \mid Z_{t-1}) = \prod_{i=1}^{N} P(Z_t^i \mid \pi(Z_t^i))$$

$Z(i,t)$ is a node at time slice t, it can be a hidden node, an observation node, or a control node(optional), $Pa(Z(i, t))$ are parent nodes of $Z(i,t)$, they can be at either time slice t or t-1. The nodes in the first slice of a 2TBN do not have parameters associated with them. But each node in the second slice has an associated CPD (conditional probability distribution).

To specify a DBN, we need to define the intra-slice topology (within a slice), the inter-slice topology (between two slices), as well as the parameters for the first two slices. (Such a two-slice temporal Bayes net is often called a 2TBN.)

We can specify the topology as follows:

```
intra = zeros(2);
intra(1,2) = 1; // node 1 in slice t connects to node 2 in slice t
inter = zeros(2);
inter(1,1) = 1; // node 1 in slice t-1 connects to node 1 in slice t
```

We can specify the parameters as follows, where for simplicity we assume the observed node is discrete.

```
Q = 2; // num hidden states
O = 2; // num observable symbols
ns = [Q O];
dnodes = 1:2;
bnet = mk_dbn(intra, inter, ns, 'discrete', dnodes);
for i=1:4 bnet.CPD{i} = tabular_CPD(bnet, i); end
eclass1 = [1 2]; eclass2 = [3 2]; eclass = [eclass1 eclass2];
bnet = mk_dbn(intra, inter, ns, 'discrete', dnodes, 'eclass1',
eclass1, 'eclass2', eclass2);
prior0 = normalise(rand(Q,1));
transmat0 = mk_stochastic(rand(Q,Q));
obsmat0 = mk_stochastic(rand(Q,O));
bnet.CPD{1} = tabular_CPD(bnet, 1, prior0);
bnet.CPD{2} = tabular_CPD(bnet, 2, obsmat0);
```

bnet.CPD{3} = tabular_CPD(bnet, 3, transmat0);

It can also be represented as:

<dbn>

  <prior>

    //...a static BN(DAG) in XMLBIF format defining the

    //state-space at time slice 1

  </prior>

  <transition>

// a transition network(DAG) including two time slices t and t+1;

    // node has an additional attribute showing which time slice it

    // belongs to only nodes in slice t+1 have CPDs

  </transition>

</dbn>

**First-order markov assumption**: the parents of a node can only be in the same time slice or the previous time slice, i.e., arcs do not across slices. Inter-slice arcs are all from left to right, reflecting the arrow of time. Intra-slice arcs can be arbitrary as long as the overall DBN is a DAG. Time-invariant assumption: the parameters of the CPDs don't change over time. The semantics of DBN can be defined by "unrolling" the 2TBN to T time slices. Fig. 11 shows the developed DBN and the resulting joint probability distribution is then defined by

$$P(Z_{1:T}) = \prod_{t=1}^{T}\prod_{i=1}^{N} P(Z_t^i \mid \pi(Z_t^i))$$

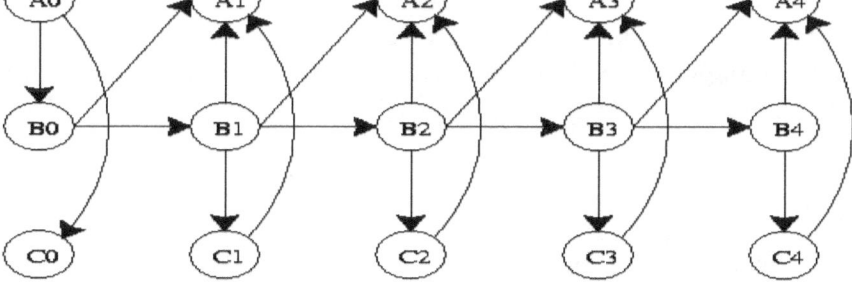

Fig. 11

# Machine Learning

## Human Learning and Machines

There have been 3 or 4 decades of research on Machine learning where scientists have tried to create a human mind in the artificial world. It is known that the ability of computer is far inferior to to that of human. As a result it seems natural to choose as a goal for learning algorithms to match human performance.

Human Brains are genetically designed for solving certain types of learning problems. There are a plenty of hard wiring of neurons done by the nature that make our visual system recognize objects easily, or hearing system comprehend sounds etc. Also, for most machine learning tasks, when we build a database of examples to be given to a computer, we only give a very limited amount of information to the computer. For example, when humans start to learn to recognize hand written characters, they already have a visual system that has been extensively trained to recognize all sorts of shapes. We imagine someone who has been blind from birth that suddenly recovers sight at the age of 6 and is directly put in front of handwritten characters on a screen, and asked to classify them. we would suspect that he would not get a better accuracy than existing learning algorithms. Another example we think of is "Spam Classification" where we think human learning is faster than machine learning.

But backtracking to the "Chapter 1" where we mentioned a point about Cognitive Bias. This is where machines can win over humans. We know machines don't have feelings, at least, yet. They won't be distracted by any external influence. We came to this consensus that Robots might never achieve that level of wisdom and humans might never leave ego.

## Definition

Machine learning, a branch of artificial intelligence, is about the construction and study of systems that can learn from data. For example, a machine learning system could be trained on email messages to learn to distinguish between spam and non-spam messages. After learning, it can then be used to classify new email messages into spam and non-spam folders.

The core of machine learning deals with representation and generalization. Representation of data instances and functions evaluated on these instances are part of all machine learning systems; for example, in the above email message example we can represent an email as a set of English words by simply discarding the word order. Generalization is the property that the system will perform well on unseen data instances; the conditions under which this can be guaranteed are a key object of study in the subfield of computational learning theory.

There is a wide variety of machine learning tasks and successful applications. Optical character recognition, in which printed characters are recognized automatically based on previous examples, is a classic engineering example of machine learning.

There are two major types of learnings:

- Supervised,
- Un-supervised

## Supervised Learning

Supervised learning is the machine learning task of inferring a function from labeled training data. The training data consist of a set of training examples. In supervised learning, each example is a pair consisting of an input object (typically a vector) and a desired output value (also called the supervisory signal). A supervised learning algorithm analyzes the training data and produces an inferred function, which is called a classifier (if the output is discrete; see classification) or a regression function (if the output is continuous; see regression). The inferred function should predict the correct output value for any valid input object. This requires the learning algorithm to generalize from the training data to unseen situations in a "reasonable" way.

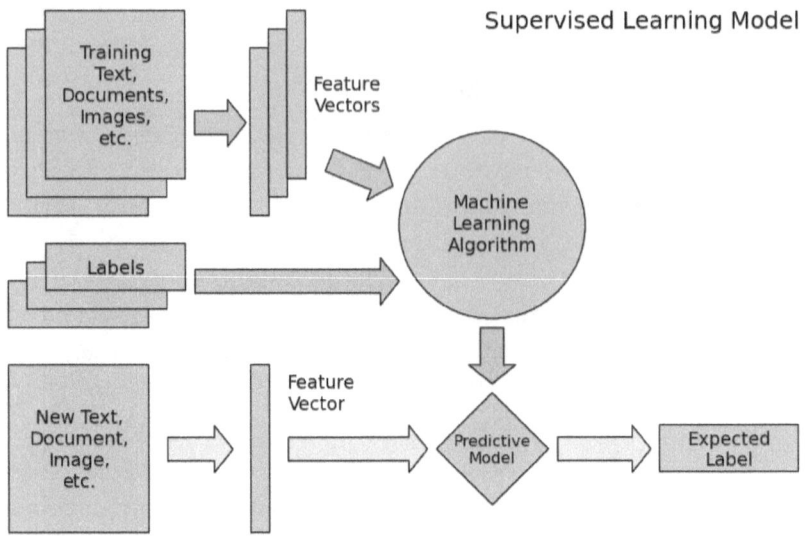

Fig. 12

There are many algorithms that are in use now a days:

- Artificial Neural Networks
- Decision Tree Learning
- Data Pre-processing
- Back propagation

## Un-Supervised Learning

In machine learning, unsupervised learning refers to the problem of trying to find hidden structure in unlabeled data. Since the examples given to the learner are unlabeled, there is no error or reward signal to evaluate a potential solution. This distinguishes unsupervised learning from supervised learning and reinforcement learning.

Unsupervised learning is closely related to the problem of density estimation in statistics. However unsupervised learning also encompasses many other techniques that seek to summarize and explain key features of the data. Many methods employed in unsupervised learning are based on data mining methods used to preprocess data.

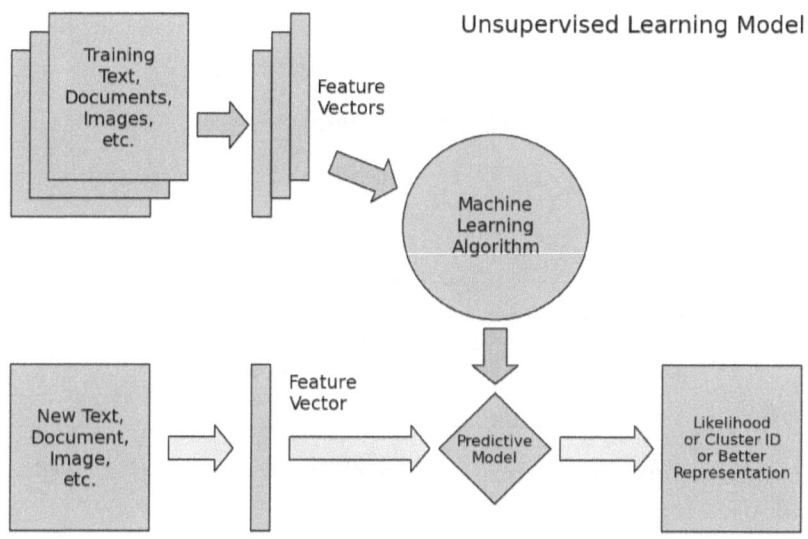

Fig. 13

Approaches to unsupervised learning include:

- Clustering (e.g., k-means, mixture models, hierarchical clustering),

- Blind signal separation using feature extraction techniques for dimensionality reduction (e.g., Principal component analysis, Independent component analysis, Non-negative matrix factorization, Singular value decomposition).

## Neural Networks

An artificial neural network (ANN), often just called a neural network (NN), is a mathematical model inspired by biological neural networks. A neural network consists of an interconnected group of artificial neurons, and it processes information using a connectionist approach to computation. In most cases a neural network is an adaptive system that changes its structure during a learning phase. Neural networks are used to model complex relationships between inputs and outputs or to find patterns in data.

Basically, a Neural Network is;

- An extremely simplified model of the brain

- Essentially a function approximator

- Transforms inputs into outputs to the best of its ability

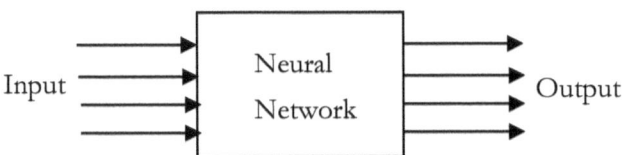

They are composed of many "neurons" that co-operate to perform the desired function.

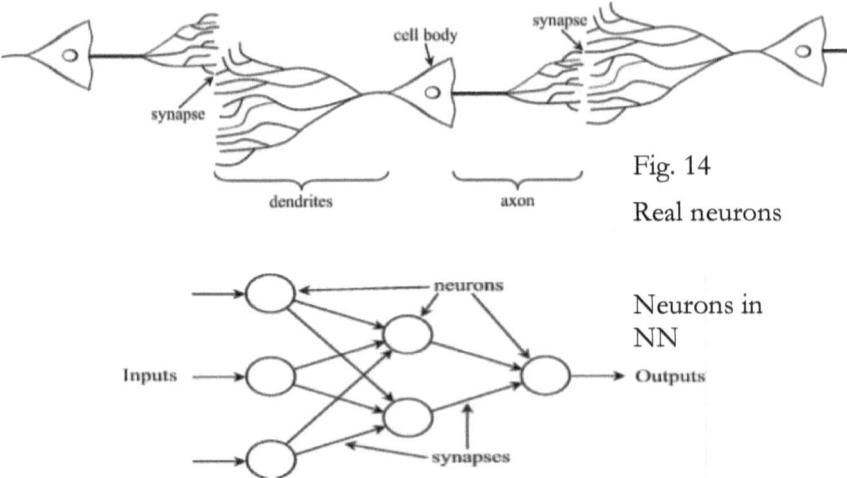

Fig. 14

Real neurons

Neurons in NN

## Basic Structure

The word network in the term 'artificial neural network' refers to the inter–connections between the neurons in the different layers of each system. An example system has three layers. The first layer has input neurons, which send data via synapses to the second layer of neurons, and then via more synapses to the third layer of output neurons. More complex systems will have more layers of neurons with some having increased layers of input neurons and output neurons. The synapses store parameters called "weights" that manipulate the data in the calculations.

An ANN is typically defined by three types of parameters:

- The interconnection pattern between different layers of neurons

- The learning process for updating the weights of the interconnections

- The activation function that converts a neuron's weighted input to its output activation.

## Why are NNs Good?

NNs are considered a good Machine Learning technique because of the following reasons:

- **Classification**: Pattern recognition, feature extraction, image matching

- **Noise Reduction**: Recognize patterns in the inputs and produce noiseless outputs

- **Prediction**: Extrapolation based on historical data

- **Ability to learn**: NN's figure out how to perform their function on their own

- Determine their function based only upon sample inputs

- **Ability to generalize**: i.e. produce reasonable outputs for inputs it has not been taught how to deal with

## Calculation of Weights

The weights in a neural network are the most important factor in determining its function.

Training is the act of presenting the network with some sample data and modifying the weights to better approximate the desired function

This can be done by two main types of training:

- Supervised Training: Supplies the neural network with inputs and the desired outputs
  - o Response of the network to the inputs is measured: The weights are modified to reduce the difference between the actual and desired outputs
- Unsupervised Training: Only supplies inputs
  - o The neural network adjusts its own weights so that similar inputs cause similar outputs
  - o The network identifies the patterns and differences in the inputs without any external assistance

## Epoch

One iteration through the process of providing the network with an input and updating the network's weights. Typically many epochs are required to train the neural network

## Mathematical Meaning

The first step in the architecture process is to define the primitive building block, and if you haven't fallen asleep at this point, you have no doubt figured out that this will be a neuron. The neuron model we will use is a version of the tried-and-true model used for software neural networks, also known as the perceptron. As illustrated in Fig4, the perceptron has multiple inputs and one output.

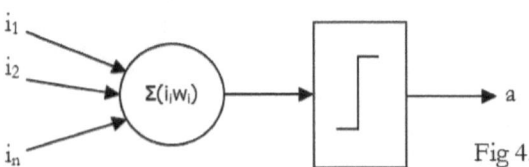

Fig 4.

The mathematical model of the perceptron is given by:

$$a = function(\Sigma(i_i w_i))$$

where,

$i_{1,2,3...n}$ are inputs,

$w_i$ is the weight to the,

'a' is the activation or output and

function(x) = 1, if x > threshold

0, otherwise

## Implementation of Universal Gates by Neural Nets
### XOR Gate

| $i_1$ | $i_2$ | a |
|---|---|---|
| 1 | 1 | 0 |
| 1 | 0 | 1 |
| 0 | 0 | 1 |
| 0 | 1 | 0 |

The network given below refers to XOR gate by using the necessary nodes and their weights.

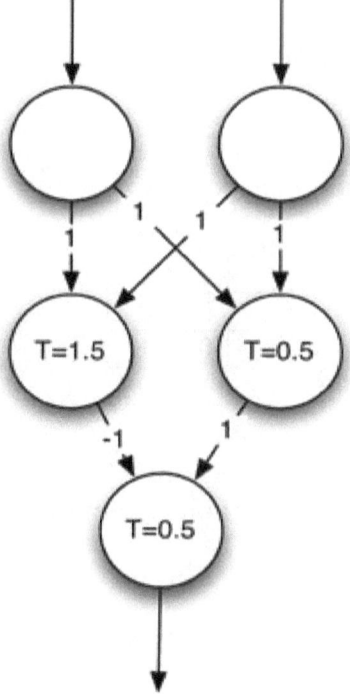

### NAND Gate

A = $i_1$, B = $i_2$

Threshold = -0.8

$w_1$, $w_2$ = -0.5

Applying the Activation Function, we get

| $i_1$ | $i_2$ | a |
|---|---|---|
| 1 | 1 | 1 |
| 1 | 0 | 1 |
| 0 | 0 | 0 |
| 0 | 1 | 1 |

**NOR Gate**

A = $i_1$, B = $i_2$

Threshold = -0.1

$w_1$, $w_2$ = -0.5

Applying the Activation Function, we get

| $i_1$ | $i_2$ | A |
|---|---|---|
| 1 | 1 | 0 |
| 1 | 0 | 0 |
| 0 | 0 | 1 |
| 0 | 1 | 0 |

Here is a code snippet for demo of NNs in C# using Perceptron learning and Sigmoid Function as thresholding function and activation network.

```
public class SigmoidFunction
{
    private double alpha = 2;
    public double Alpha
```

```
    {
        get { return alpha; }
        set { alpha = value; }
    }
    public SigmoidFunction( ) { }
    public SigmoidFunction( double alpha ){
        this.alpha = alpha;
    }
    public double Function( double x ){
        return ( 1 / ( 1 + Math.Exp( -alpha * x
) ) );
    }
    public double Derivative( double x ){
        double y = Function( x );
        return ( alpha * y * ( 1 - y ) );
    }
    public double Derivative2( double y ){
        return ( alpha * y * ( 1 - y ) );
    }}


public ActivationNetwork( IActivationFunction
function, int inputsCount, params int[]
neuronsCount )
                                        : base(
inputsCount, neuronsCount.Length )
{
    // create each layer
    for ( int i = 0; i < layersCount; i++ )
    {
        layers[i] = new ActivationLayer(
            // neurons count in the layer
            neuronsCount[i],
            // inputs count of the layer
            ( i == 0 ) ? inputsCount :
neuronsCount[i - 1],
            // activation function of the
layer
```

```
                        function );
    }
}


public ActivationLayer( int neuronsCount, int
inputsCount, IActivationFunction function )
                            : base( neuronsCount,
inputsCount ){
    for ( int i = 0; i < neuronsCount; i++ )
        neurons[i] = new ActivationNeuron(
inputsCount, function );
}


public class PerceptronLearning :
ISupervisedLearning
{
    // network to teach
    private ActivationNetwork network;
    // learning rate

    private double learningRate = 0.1;
    public double LearningRate
    {
        get { return learningRate; }
        set
        {
            learningRate = Math.Max( 0.0,
Math.Min( 1.0, value ) );
        }
    }
    public PerceptronLearning( ActivationNetwork
network )
    {
        if ( network.LayersCount != 1 )
        {
            throw new ArgumentException(
"Invalid nuaral network. It should have one layer
only." );
```

```
        }
        this.network = network;
    }
    public double Run( double[] input, double[]
output )
    {
        double[] networkOutput =
network.Compute( input );
        ActivationLayer layer = network[0];
        double error = 0.0;
        for ( int j = 0, k = layer.NeuronsCount;
j < k; j++ )
        {
            double e = output[j] -
networkOutput[j];
            if ( e != 0 )
            {
                ActivationNeuron perceptron
= layer[j];
                for ( int i = 0, n =
perceptron.InputsCount; i < n; i++ )
                {
                    perceptron[i] +=
learningRate * e * input[i];
                }
                perceptron.Threshold +=
learningRate * e;

                error += Math.Abs( e );
            }
        }

        return error;
    }
public double RunEpoch( double[][] input,
double[][] output )
{
double error = 0.0;
for ( int i = 0, n = input.Length; i < n; i++)
```

```
{       error += Run( input[i], output[i] );}
        return error;
}}
```

This code clearly demonstrates the working of neural networks in the field of learning in AI.

# Reinforcement Learning

Inspired by behaviorist psychology, reinforcement learning is an area of machine learning in computer science, concerned with how an agent ought to take actions in an environment so as to maximize some notion of cumulative reward. The problem, due to its generality, is studied in many other disciplines, such as game theory, control theory, operations research, information theory, simulation-based optimization, statistics, and genetic algorithms. In the operations research and control literature the field where reinforcement learning methods are studied is called approximate dynamic programming. The problem has been studied in the theory of optimal control, though most studies there are concerned with existence of optimal solutions and their characterization, and not with the learning or approximation aspects. In economics and game theory, reinforcement learning may be used to explain how equilibrium may arise under bounded rationality.

In machine learning, the environment is typically formulated as a Markov decision process (MDP), and many reinforcement learning algorithms for this context are highly related to dynamic programming techniques. The main difference to these classical techniques is that reinforcement learning algorithms do not need the knowledge of the MDP and they target large MDPs where exact methods become infeasible.

Reinforcement learning differs from standard supervised learning in that correct input/output pairs are never presented, nor sub-optimal actions explicitly corrected. Further, there is a focus on on-line performance, which involves finding a balance between exploration (of uncharted territory) and exploitation (of current knowledge). The exploration vs. exploitation trade-off in reinforcement learning has been most thoroughly studied through the multi-armed bandit problem and in finite MDPs.

## Markov Decision Process

A reinforcement learning task that satisfies the Markov property is called a Markov decision process, or MDP. If the state and action spaces are finite, then it is called a finite Markov decision process (finite MDP). Finite MDPs are particularly important to the theory of reinforcement learning. A particular finite MDP is defined by its state and action sets and by the one-step dynamics of the environment. Given any state and action, $s$ and $a$, the probability of each possible next state, $s'$, is

$$\mathcal{P}^a_{ss'} = Pr\left\{s_{t+1} = s' \mid s_t = s, a_t = a\right\}.$$

These quantities are called *transition probabilities*. Similarly, given any current state and action, $s$ and $a$, together with any next state, $s'$, the expected value of the next reward is

$$\mathcal{R}^a_{ss'} = E\left\{r_{t+1} \mid s_t = s, a_t = a, s_{t+1} = s'\right\}.$$

These quantities, $\mathcal{P}^a_{ss'}$ and $\mathcal{R}^a_{ss'}$, completely specify the most important aspects of the dynamics of a finite MDP (only information about the distribution of rewards around the expected value is lost). Most of the theory we present in the rest of this book implicitly assumes the environment is a finite MDP.

**Example (Recycling Robot MDP):** This can be used as a simple example of an MDP by simplifying it and providing some more details. (Our aim is to produce a simple example, not a particularly realistic one.) Recall that the agent makes a decision at times determined by external events (or by other parts of the robot's control system). At each such time the robot decides whether it should (1) actively search for a can, (2) remain stationary and wait for someone to bring it a can, or (3) go back to home base to recharge its battery. Suppose the environment works as follows. The best way to find cans is to actively search for them, but this runs down the robot's battery, whereas waiting does not. Whenever the robot is searching, the possibility exists that its battery will become depleted. In this case the robot must shut down and wait to be rescued (producing a low reward).

The agent makes its decisions solely as a function of the energy level of the battery. It can distinguish two levels, high and low, so that the state set is $S = \{\text{high, low}\}$. Let us call the possible decisions--the agent's actions--wait, search, and recharge. When the energy level is high, recharging would always be foolish, so we do not include it in the action set for this state. The agent's action sets are

$$\mathcal{A}(\text{high}) = \{\text{search, wait}\}$$
$$\mathcal{A}(\text{low}) = \{\text{search, wait, recharge}\}.$$

If the energy level is high, then a period of active search can always be completed without risk of depleting the battery. A period of searching that begins with a high energy level leaves the energy level high with probability $\alpha$ and reduces it to low with probability $1 - \alpha$. On the other hand, a period of searching undertaken when the energy level is low leaves it low with probability $\beta$ and depletes the battery with probability $1 - \beta$. In the latter case, the robot must be rescued, and the battery is then recharged back to high. Each can collected by the robot counts as a unit reward, whereas a reward of $-3$ results whenever the robot has to be rescued. Let $\mathcal{R}^{\text{search}}$ and $\mathcal{R}^{\text{wait}}$, with $\mathcal{R}^{\text{search}} > \mathcal{R}^{\text{wait}}$, respectively denote the expected number of cans the robot will collect (and hence the expected reward) while searching and while waiting. Finally, to keep things simple, suppose that no cans can be collected during a run home for recharging, and that no cans can be collected on a step in which the battery is depleted. This system is then a finite MDP, and we can write down the transition probabilities and the expected rewards, as in Fig.15. A transition graph is a useful way to summarize the dynamics of a finite MDP. Fig.16 shows the transition graph for the recycling robot example.

There are two kinds of nodes: state nodes and action nodes. There is a state node for each possible state (a large open circle labeled by the name of the state), and an action node for each state-action pair (a small solid circle labeled by the name of the action and connected by a line to the state node). Starting in state $s$ and taking action $a$ moves you along the line from state node $s$ to action node $(s, a)$. Then the environment responds with a transition to the next state's node via one of the arrows leaving action node $(s, a)$. Each arrow corresponds to a triple $(s, s', a)$, where $s'$ is the next state, and we label the arrow with the transition probability, $\mathcal{P}^a_{ss'}$, and the expected reward for that transition, $\mathcal{R}^a_{ss'}$. Note that the transition probabilities labeling the arrows leaving an action node always sum to 1.

| $s = s_t$ | $s' = s_{t+1}$ | $a = a_t$ | $\mathcal{P}^a_{ss'}$ | $\mathcal{R}^a_{ss'}$ |
|---|---|---|---|---|
| high | high | search | $\alpha$ | $\mathcal{R}^{\text{search}}$ |
| high | low | search | $1 - \alpha$ | $\mathcal{R}^{\text{search}}$ |
| low | high | search | $1 - \beta$ | $-3$ |
| low | low | search | $\beta$ | $\mathcal{R}^{\text{search}}$ |
| high | high | wait | $1$ | $\mathcal{R}^{\text{wait}}$ |
| high | low | wait | $0$ | $\mathcal{R}^{\text{wait}}$ |
| low | high | wait | $0$ | $\mathcal{R}^{\text{wait}}$ |
| low | low | wait | $1$ | $\mathcal{R}^{\text{wait}}$ |
| low | high | recharge | $1$ | $0$ |
| low | low | recharge | $0$ | $0.$ |

Fig. 15

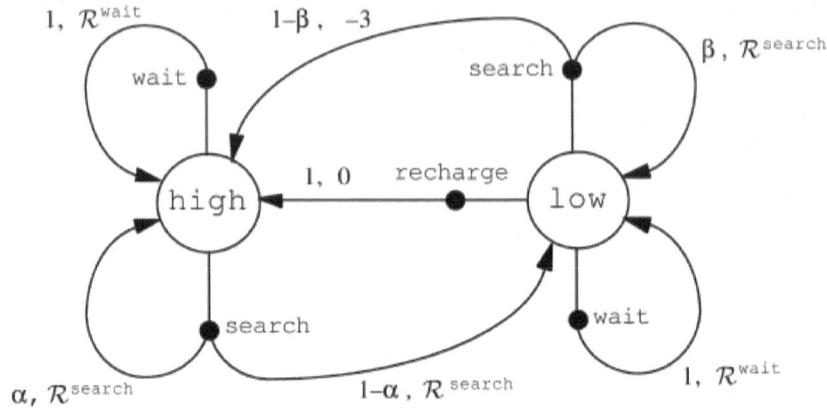

Fig. 16

## POMDP (Partially observable MDP)

A POMDP is really just an MDP; we have a set of states, a set of actions, transitions and immediate rewards. The actions' effects on the state in a POMDP are exactly the same as in an MDP. The only difference is in whether or not we can observe the current state of the process. In a POMDP we add a set of observations to the model. So instead of directly observing the current state, the state gives us an observation which provides a hint about what state it is in. The observations can be probabilistic; so we need to also specify the observation model. This observation model simply tells us the probability of each observation for each state in the model.

Although the underlying dynamics of the POMDP are still Markovian, since we have no direct access to the current state, our decisions require keeping track of (possibly) the entire history of the process, making this a non-Markovian process. The history at a given point in time is comprised of our knowledge about our starting situation, all actions performed and all observations seen. A POMDP models an agent decision process in which it is assumed that the system dynamics are determined by an MDP, but the agent cannot directly observe the underlying state. Instead, it must maintain a probability distribution over the set of possible states, based on a set of observations and observation probabilities, and the underlying MDP.

The POMDP framework is general enough to model a variety of real-world sequential decision processes. Applications include robot navigation problems, machine maintenance, and planning under uncertainty in general. The framework originated in the Operations Research community, and was later taken over by the Artificial Intelligence and Automated Planning communities. An exact solution to a POMDP yields the optimal action for each possible belief over the world states. The optimal action maximizes (or minimizes) the expected reward (or cost) of the agent over a possibly infinite horizon. The sequence of optimal actions is known as the optimal policy of the agent for interacting with its environment.

The advanced algorithms and models of artificial intelligence dedicated to imitate humans at the farthest level will be discussed in the next chapter which will include topics such as computer vision, natural language processing and robotics.

# Chapter 4

# Working on Advanced Applications

## Natural Language Processing

### Human Language Processing

#### Psycholinguistics

"To understand and model the processes that underlies the human capacity to understand language"

- How does the human language processor work?

- How is it realized in the brain?

- How is linguistic knowledge represented in the brain?

- How can we understanding computationally?

- Where does our capacity for language emerge from?

The brain is the natural computer, par excellence where perception occurs in real time, and is highly strategic

Traditional views on human perception:

- Cognitivist: inferential, unencapsulated: cognitive penetration of perceptual processes.

- Behaviourist: non-inferential, encapsulated: perception reduces to conditioned reflexes.

## Functions

What does it do?

- Comprehension: Maps from "sound to meaning"
    - speech/orthography to words
    - words to structures
    - structure to meanings
- Production: Maps from "message to speech"
    - Meaning to grammatical encoding
    - Phonological encoding
    - Articulation

## Competence vs Knowledge

Competence: Knowledge of Language

- Linguistic theories at all levels: Phonetics/phonology, morphology, syntax, semantics ...

- Rules and representations

Performance: How Language is Processing

- Use of Knowledge of Language

- Processes for comprehension and production

- Architectures and Mechanisms

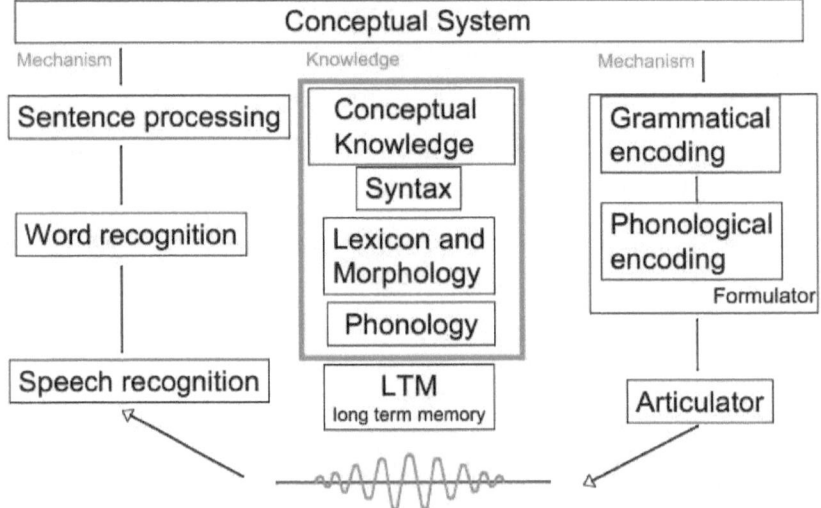

Speech Processing Model (Dijkstra and Kempen, 1993)

## Relation with NLP

NLP is a form of human-to-computer interaction where the elements of human language, be it spoken or written, are formalized so that a computer can perform value-adding tasks based on that interaction.

Natural language understanding has captivated human imagination throughout generations. A central issue is how to discover and connect various meaningful concepts from human-generated content, such as text, speech and media. A key objective is to resolve the inherent ambiguities in natural languages.

Natural languages are those spoken by people.

NLP encompasses anything a computer needs to understand natural language (typed or spoken) and also generate the natural language.

Natural Language Processing (NLP) is a subfield of Artificial intelligence and linguistics, devoted to make computers "understand" statements written in human languages.

NLP encompasses anything a computer needs to understand natural language (typed or spoken) and also generate the natural language.

**Natural Language Understanding (NLU)**

The NLU task is "understanding and reasoning" while the input is a natural language. Here we ignore the issues of natural language generation.

**Natural Language Generation (NLG)**

NLG is a subfield of natural language processing NLP. NLG is also referred to text generation.

# Steps of Natural Language Processing (NLP)

Natural Language Processing is done at 5 levels, as shown in the previous slide. These levels are briefly stated below.

- **Morphological and Lexical Analysis**: The lexicon of a language is its vocabulary that includes its words and expressions. Morphology is the identification, analysis and description of structure of words. The words are generally accepted as being the smallest units of syntax. The syntax refers to the rules and principles that govern the sentence structure of any individual language.

  The aim is to divide the text into paragraphs, Lexical analysis: sentences and words. the lexical analysis cannot be performed in isolation from morphological and syntactic analysis

- **Syntactic Analysis**: Here the analysis is of words in a sentence to know the grammatical structure of the sentence.

The words are transformed into structures that show how the words relate to each others. Some word sequences may be rejected if they violate the rules of the language for how words may be combined.

Example: An English syntactic analyzer would reject the sentence say:

"Boy the go the to store".

- **Semantic Analysis**: It derives an absolute (dictionary definition) meaning from context; it determines the possible meanings of a sentence in a context. The structures created by the syntactic analyzer are assigned meaning.

  Thus, a mapping is made between the syntactic structures and objects in the task domain. The structures for which no such mapping is possible are rejected.

  Example: the sentence would be rejected "Colorless green ideas . . . "as semantically anomalous because colorless and green make no sense.

- **Discourse Integration**: The meaning of an individual sentence may depend on the sentences that precede  it and may influence the meaning of the sentences that follow it.

  Example: the word in the sentence, depends  " it " , "you wanted it" on  the  prior  discourse  context.

- **Pragmatic analysis**: It derives knowledge from external commonsense  information; it means understanding the purposeful use of language in situations, particularly those aspects of language which require world knowledge; the idea is, what was said is reinterpreted to determine what was actually meant.

  Example:  the sentence "Do you know what time it is?" should be interpreted as a request.

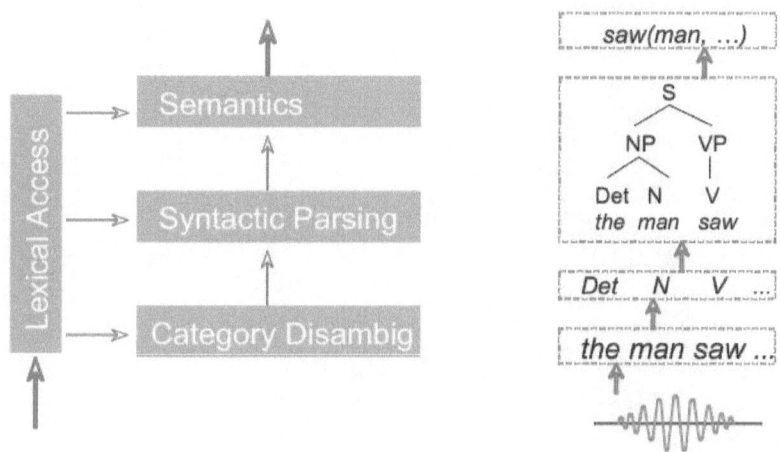

The NLP Model Architecture

## POS Tagger

In corpus linguistics, part-of-speech tagging (POS tagging or POST), also called grammatical tagging or word-category disambiguation, is the process of marking up a word in a text (corpus) as corresponding to a particular part of speech, based on both its definition, as well as its context—i.e. relationship with adjacent and related words in a phrase, sentence, or paragraph. A simplified form of this is commonly taught to school-age children, in the identification of words as nouns, verbs, adjectives, adverbs, etc.

```
public class MaximumEntropyPosTagger
{
    private SharpEntropy.IMaximumEntropyModel
mPosModel;
    private IPosContextGenerator
mContextGenerator;
    private PosLookupList mDictionary;
    private bool mUseClosedClassTagsFilter =
false;
    private const int mDefaultBeamSize = 3;
    private int mBeamSize;
    private Util.Sequence mBestSequence;
    public virtual string NegativeOutcome{
        get{return "";}
```

```
        }
    public virtual int NumTags{
        get{return mPosModel.OutcomeCount;}
    }
    public virtual string[] AllTags(){
        string[] tags = new
string[mPosModel.OutcomeCount];
        for (int currentTag = 0; currentTag <
mPosModel.OutcomeCount; currentTag++){
            tags[currentTag] =
mPosModel.GetOutcomeName(currentTag);
        }
        return tags;}
    protected internal
SharpEntropy.IMaximumEntropyModel PosModel{
        get{return mPosModel;}
        set{mPosModel = value;}
    }
    protected internal IPosContextGenerator
ContextGenerator{
        get{return mContextGenerator;}
        set{mContextGenerator = value;}
    }
    protected internal PosLookupList
TagDictionary{
        get{return mDictionary;}
        set{mDictionary = value;}
    }
    protected internal bool
UseClosedClassTagsFilter{
        get{return mUseClosedClassTagsFilter;}
        set{mUseClosedClassTagsFilter = value;}
    }
    protected internal int BeamSize{
        get{return mBeamSize;}
        set{mBeamSize = value;}
    }
    internal Util.BeamSearch Beam;
    public
MaximumEntropyPosTagger(SharpEntropy.IMaximumEntro
pyModel model) : this(model, new
DefaultPosContextGenerator()){}
    public
MaximumEntropyPosTagger(SharpEntropy.IMaximumEntro
pyModel model, PosLookupList dictionary) :
this(mDefaultBeamSize, model, new
DefaultPosContextGenerator(), dictionary){}
    public
MaximumEntropyPosTagger(SharpEntropy.IMaximumEntro
pyModel model, IPosContextGenerator
contextGenerator) : this(mDefaultBeamSize, model,
contextGenerator, null){}
```

```
     public
MaximumEntropyPosTagger(SharpEntropy.IMaximumEntro
pyModel model, IPosContextGenerator
contextGenerator, PosLookupList dictionary) :
this(mDefaultBeamSize, model, contextGenerator,
dictionary){}
     public MaximumEntropyPosTagger(int beamSize,
SharpEntropy.IMaximumEntropyModel model,
IPosContextGenerator contextGenerator,
PosLookupList dictionary){
          mBeamSize = beamSize;
          mPosModel = model;
          mContextGenerator = contextGenerator;
          Beam = new PosBeamSearch(this,
mBeamSize, contextGenerator, model);
          mDictionary = dictionary;
     }
     public virtual
SharpEntropy.ITrainingEventReader
GetEventReader(System.IO.TextReader reader){
          return new PosEventReader(reader,
mContextGenerator);
     }
     public virtual ArrayList Tag(ArrayList tokens)
{
          mBestSequence =
Beam.BestSequence(tokens, null);
          return new
ArrayList(mBestSequence.Outcomes);
     }
     public virtual string[] Tag(string[] tokens) {
          mBestSequence = Beam.BestSequence(new
ArrayList(tokens), null);
          return
mBestSequence.Outcomes.ToArray();
     }
     public virtual void GetProbabilities(double[]
probabilities) {

     mBestSequence.GetProbabilities(probabilities);
     }
     public virtual double[] GetProbabilities(){
          return mBestSequence.GetProbabilities();
     }
     public virtual string TagSentence(string
sentence)
     {
          ArrayList tokens = new
ArrayList(sentence.Split());
          ArrayList tags = Tag(tokens);
          System.Text.StringBuilder tagBuffer =
new System.Text.StringBuilder();
```

```csharp
        for (int currentTag = 0; currentTag <
tags.Count; currentTag++)

    tagBuffer.Append(tokens[currentTag] + "/" +
tags[currentTag] + " ");
        return tagBuffer.ToString().Trim();
    }
    public virtual void
LocalEvaluate(SharpEntropy.IMaximumEntropyModel
posModel, System.IO.StreamReader reader, out
double accuracy, out double sentenceAccuracy)
    {
        mPosModel = posModel;
        float total = 0, correct = 0, sentences
= 0, sentencesCorrect = 0;
        System.IO.StreamReader sentenceReader =
new System.IO.StreamReader(reader.BaseStream,
System.Text.Encoding.UTF7);
        string line;
        while ((object) (line =
sentenceReader.ReadLine()) != null) {
            sentences++;
            Util.Pair<ArrayList, ArrayList>
annotatedPair =
PosEventReader.ConvertAnnotatedString(line);
            ArrayList words =
annotatedPair.FirstValue;
            ArrayList outcomes =
annotatedPair.SecondValue;
            ArrayList tags = new
ArrayList(Beam.BestSequence(words,
null).Outcomes);

            int count = 0;
            bool isSentenceOK = true;
            for
(System.Collections.IEnumerator tagIndex =
tags.GetEnumerator(); tagIndex.MoveNext();
count++) {
                total++;
                string tag = (string)
tagIndex.Current;
                if (tag ==
(string)outcomes[count]) {
                    correct++;
                }
                else {
                    isSentenceOK = false;
                }
            }
            if (isSentenceOK) {
                sentencesCorrect++;
            }
```

```
        }
        accuracy = correct / total;
        sentenceAccuracy = sentencesCorrect /
sentences;
    }
    private class PosBeamSearch : Util.BeamSearch
{
        private MaximumEntropyPosTagger
mMaxentPosTagger;
        public
PosBeamSearch(MaximumEntropyPosTagger posTagger,
int size, IPosContextGenerator contextGenerator,
SharpEntropy.IMaximumEntropyModel model) :
base(size, contextGenerator, model) {
            mMaxentPosTagger = posTagger;
        }
        public
PosBeamSearch(MaximumEntropyPosTagger posTagger,
int size, IPosContextGenerator contextGenerator,
SharpEntropy.IMaximumEntropyModel model, int
cacheSize) : base(size, contextGenerator, model,
cacheSize) {
            mMaxentPosTagger = posTagger;
        }
        protected internal override bool
ValidSequence(int index, object[] inputSequence,
string[] outcomesSequence, string outcome)  {
            if (mMaxentPosTagger.TagDictionary
== null)  {return true;}
            else {
                string[] tags =
mMaxentPosTagger.TagDictionary.GetTags(inputSequen
ce[index].ToString());
                if (tags == null) {return
true;}
                else  {return new
ArrayList(tags).Contains(outcome);}
            }
        }
        protected internal override bool
ValidSequence(int index, ArrayList inputSequence,
Util.Sequence outcomesSequence, string outcome) {
            if (mMaxentPosTagger.mDictionary
== null){return true;}
            else{
                string[] tags =
mMaxentPosTagger.mDictionary.GetTags(inputSequence
[index].ToString());
                if (tags == null){return
true;}
                else{return new
ArrayList(tags).Contains(outcome);}
            }
```

```
        }
    }
    public virtual string[]
GetOrderedTags(ArrayList words, ArrayList tags,
int index){
        return GetOrderedTags(words, tags,
index, null);
    }
    public virtual string[]
GetOrderedTags(ArrayList words, ArrayList tags,
int index, double[] tagProbabilities) {
        double[] probabilities =
mPosModel.Evaluate(mContextGenerator.GetContext(in
dex, words.ToArray(), (string[])
tags.ToArray(typeof(string)), null));
        string[] orderedTags = new
string[probabilities.Length];
        for (int currentProbability = 0;
currentProbability < probabilities.Length;
currentProbability++) {
            int max = 0;
            for (int tagIndex = 1; tagIndex <
probabilities.Length; tagIndex++) {
                if (probabilities[tagIndex]
> probabilities[max]) {max = tagIndex;}
            }
            orderedTags[currentProbability] =
mPosModel.GetOutcomeName(max);
            if (tagProbabilities != null)
{tagProbabilities[currentProbability] =
probabilities[max];}
            probabilities[max] = 0;
        }
        return orderedTags;
    }
    public static SharpEntropy.GisModel
Train(SharpEntropy.ITrainingEventReader
eventStream, int iterations, int cut){
        SharpEntropy.GisTrainer trainer = new
SharpEntropy.GisTrainer();
        trainer.TrainModel(iterations, new
SharpEntropy.TwoPassDataIndexer(eventStream,
cut));
        return new
SharpEntropy.GisModel(trainer);
    }
    public static SharpEntropy.GisModel
TrainModel(string trainingFile){
        return TrainModel(trainingFile, 100, 5);
    }
    public static SharpEntropy.GisModel
TrainModel(string trainingFile, int iterations,
int cutoff){
```

```
        SharpEntropy.ITrainingEventReader
eventReader = new PosEventReader(new
System.IO.StreamReader(trainingFile));
        return Train(eventReader, iterations,
cutoff);
    }
}
```

## Parser

It parses the while text and tries to identify the sentences and divide into its constituents.

```
/*Returns the probability associated with the
pos-tag sequence assigned to this parse.
*/
public virtual double GetTagSequenceProbability()
{
if (mParts.Count == 1 && (mParts[0]).Type ==
MaximumEntropyParser.TokenNode){
    return System.Math.Log(mProbability);
}
else{
if (mParts.Count == 0){
    throw new
ParseException("Parse.GetTagSequenceProbability()
: Wrong base case!");
}
else{
    double sum = 0.0;
    foreach (Parse oChildParse in mParts){
        sum +=
oChildParse.GetTagSequenceProbability();
    }
    return sum;
}}}

/* Returns the parse nodes which are children of
this node and which are pos tags.
*/
public virtual Parse[] GetTagNodes(){
List<Parse> tags = new List<Parse>();
List<Parse> nodes = new List<Parse>(mParts);
while (nodes.Count != 0){
Parse currentParse = nodes[0];
nodes.RemoveAt(0);
if (currentParse.IsPosTag){
    tags.Add(currentParse);
}
else{
    nodes.InsertRange(0,
currentParse.GetChildren());
```

```
}}
return tags.ToArray();
}

/* The pattern used to find the base constituent
label of a Penn Treebank labeled constituent.
*/
private static Regex mTokenPattern = new
Regex("^[^ ()]+ ([^ ()]+)\\s*\\)");
rivate static string GetType(string rest)

if (rest.StartsWith("-LCB-")){return "-LCB-";}
else if (rest.StartsWith("-RCB-")){return "-RCB-
";}
else if (rest.StartsWith("-LRB-")){return "-LRB-
";}
else if (rest.StartsWith("-RRB-")){return "-RRB-
";}
else{
MatchCollection typeMatches =
mTypePattern.Matches(rest);
if (typeMatches.Count > 0){
   return typeMatches[0].Value;
}}
return null;
}
```

## Tokenizer

Given a piece of text, we want to split the text at all spaces (including new line characters and carriage returns) and punctuation marks.

The algorithm works as follows:

- Modify the Greedy Tokenizer user interface by adding a group box containing two radio buttons for the user to decide how to deal with digits.
- Create two different splitting character arrays. One has the digit characters which is used when the user decides to throw away numbers. The other does not have the digits which is used when the user decides to keep the numbers. But none of them contains the "'" (apostrophe) character.
- Split the text into arrays using a particular separator based on the check-state of the radio button.
- Examine each resulted token discriminatively. See the code comment for details.

```csharp
// Use this to keep digits.
private static char[] delimiters_keep_digits =
new char[] {
'{', '}', '(', ')', '[', ']', '>',
'<', '-', '=', '+',
'~', '|', '\\', ':', ';', ' ', ',', '.', '/', '?',
'@', '#', '$', '%', '^', '&', '*', ' ', '\r',
'\n', '\t' };
// This will discard digits
private static char[] delimiters_no_digits = new
char[] {
'{', '}', '(', ')', '[', ']', '>',
'<', '-', '=', '+',
'~', '|', '\\', ':', ';', ' ', ',', '.', '/', '?',
'@', '#', '$', '%', '^', '&', '*', ' ', '\r',
'\n', '\t',
'0', '1', '2', '3', '4', '5', '6', '7', '8',
'9' };
//  Tokenizes a text into an array of words,
using whitespace and
/*  all punctuation except the apostrophe "'" as
delimiters. Digits
re handled based on user choice.
*/
tokens</returns>
public static string[] Tokenize( string text,
bool keepDigits ){
    string[] tokens = null;
    if ( keepDigits )
        tokens = text.Split(
delimiters_keep_digits,
StringSplitOptions.RemoveEmptyEntries );
    else
        tokens = text.Split(
delimiters_no_digits,
StringSplitOptions.RemoveEmptyEntries );
    for ( int i = 0; i < tokens.Length; i++ )
    {
        string token = tokens [i];
        // Change token only when it starts
and/or ends with "'" and the
        // toekn has at least 2 characters.
        if ( token.Length > 1 )
        {
            if ( token.StartsWith( "'" ) &&
token.EndsWith( "'" ) )
```

```
            tokens [i] = token.Substring( 1,
token.Length - 2 ); // remove the starting and
ending "'"

            else if ( token.StartsWith( "'" ) )
                tokens [i] = token.Substring( 1
); // remove the starting "'"

            else if ( token.EndsWith( "'" ) )
                tokens [i] = token.Substring( 0,
token.Length - 1 ); // remove the last "'"
            }
        }
        return tokens; }}
```

## Average type token ratio

As any language's vocabulary is finite but the length of a text can be infinitely long since we can, in theory, at least, append texts to texts an infinite number of times, type token ratio is affected by a text's length. In general, the longer a text, the smaller its type token ratio. We want to overcome this problem by computing the average type token ratio of a text independent of its length.

```
//  Computes the average type token ratio of an
array of tokens, based on the window size.
private static double GetAverageTypeTokenRatio(
string[] tokens, int windowSize ){
    LinkedList<string> movingWindow = new
LinkedList<string>( );
    int index = 0;
    while ( index < windowSize ){
        movingWindow.AddLast( tokens [index] );
        index++;
    }
    // Build frequency table of this window of
tokens
    Dictionary<string, int> movingFreqTable =
BuildFreqTable( movingWindow );
    // This type token ratio keeps changing
    double finalTTR = ( double
)movingFreqTable.Count / movingWindow.Count;
    int windowCount = 1;
    // Now index stops at windowSize position of
the tokens.
    while ( index < tokens.Length )
    {
```

```
                // Check the first token of the moving
window of tokens and remove it from
                // the moving window.
        string firstToken =
movingWindow.First.Value;
        movingWindow.RemoveFirst( );

                // Check its frequency in the frequency
table. If it is 1, it means that this token
                // occurs in the moving window only once,
so we can safely remove it from the moving
                // frequency table; otherwise, it appears
more than once, so we cannot delete it but
                // we can reduce its frequency by 1.
        if ( movingFreqTable [firstToken] == 1 )
            movingFreqTable.Remove( firstToken );
        else
            movingFreqTable [firstToken]--;

                // Find the next available token. If it is
in the moving frequency table, increase its
                // frequency value by 1; otherwise, add it
as a new entry and set its frequency to 1.
        string newToken = tokens [index];

        if ( movingFreqTable.ContainsKey( newToken
) )
            movingFreqTable [newToken]++;
        else
            movingFreqTable.Add( newToken, 1 );

                // Add this token to the moving window so
that the window always has the same number of
tokens.
        movingWindow.AddLast( newToken );

                // Re-compute the type token ratio of this
changed window.
        double thisTTR = ( double
)movingFreqTable.Count / windowSize;

                // Add this new type token ratio to the
final type token ratio.
        finalTTR += thisTTR;

                // Update index position and window
counters
        index++;
        windowCount++; }

    finalTTR = finalTTR / windowCount;

    return finalTTR;
```

```
}

// Compute the general type token ratio of an
array of tokens.
private static void GetGeneralTypeTokenRatio(
string[] tokens,
                    out HashSet<string> types,
out double typeTokenRatio )
{
    // Dump array of tokens into a HashSet of
string. By definition,
    // HashSet has no duplicates, which is the
what type means.
    types = new HashSet<string>( );

    foreach ( string token in tokens )
    types.Add( token );

    // A sanity check: if types set is empty, set
//typeTokenRatio = double.NaN,

  if ( types.Count == 0 )
    {
        typeTokenRatio = double.NaN;
    }
    else
    {
        // Be very aware that you need to cast
//either types.Count or tokens.Length

        typeTokenRatio = ( double )types.Count /
tokens.Length;
    }}
// Create a string-integer dictionary out of a
linked list of tokens.
private static Dictionary<string, int>
BuildFreqTable( LinkedList<string> tokens )
{
Dictionary<string, int> token_freq_table = new
Dictionary<string, int>( );
foreach ( string token in tokens )
{
    if ( token_freq_table.ContainsKey( token ) )
        token_freq_table [token]++;
    else
        token_freq_table.Add( token, 1 );
}
return token_freq_table; }
```

These are some of the codes developed to show the
implementation of important NLP components.

# Human Perception and Computer Vision

## Is the human visual perception good enough?

How many distinct line lengths and orientations can humans accurately perceive? How many different sound pitches or volumes can we distinguish without error? What is our "channel capacity" when dealing with color, taste, smell, or any other of our senses? How are humans capable of recognizing hundreds of faces and thousands of spoken words?

These and related issues are important in the study of both computer vision and scientific visualization. On the one hand, attempting to identify the limits of human perception can lead to insights into the design of image understanding systems. On the other hand, when designing visualization it is important to factor in these limitations to avoid generating images with ambiguous or misleading information. This talk will present an overview of some very early work on perceptual psychology and relate it to current work in image science.

## Image Formation

The images we process in computer vision are formed by light bouncing off surfaces in the world and into the lens of the system. The light then hits an array of sensors inside the camera. Each sensor produces electric charges that are read by an electronic circuit and converted to voltages. These are in turn sampled by a device called a digitizer (or analog-to-digital converter) to produce the numbers that computers eventually process, called pixel values. Thus, the pixel values are a rather indirect encoding of the physical properties of visible surfaces. Is it not amazing that all those numbers in an image file carry information on how the properties of a packet of photons were changed by bouncing off a surface in the world? Even more amazing is that from this information we can perceive shapes and colors. Although we are used to these notions nowadays, the discovery of how images form, say, on our retinas, is rather recent. In ancient Greece, Euclid, in 300 B.C., attributed sight to the action of rectilinear rays issuing from the observer's eye, a theory that remained prevalent until the sixteenth Century when Johannes Kepler explained image formation as we understand it now. In Euclid's view, then, the eye is an active participant in the visual process. Not a receptor, but an agent that reaches out to apprehend its object. One of Euclid's postulates on vision maintained that any given object can be removed to a distance from which it will no longer be visible because it falls between adjacent visual rays. This is ray tracing in a very concrete, physical sense!

Studying image formation amounts to formulating models of the process that encodes the properties of light off a surface in the world into brightness values in the image array. We start from what happens once light leaves a visible surface. What happens thereafter is in fact a function only of the imaging device, if we assume that the medium in-between is transparent. In contrast, what happens at the visible surface, although definitely of great interest in computer vision is so to speak out of our control, because it depends on the reflectance properties of the surface. In other words, reflectance is about the world, not about the imaging process. The study of image formation can be further divided into what happens up to the point when light hits the sensor, and what happens thereafter. The first part occurs in the realm of optics, the second is a matter of electronics.

## Image Science

Visual perception is a function of our eyes and brain. We see images as a whole rather then in parts. However, images can be broken down into their visual elements: line, shape, texture, and color. These elements are to images as grammar is to language. Together they allow our eyes to see images and our brain to recognize them. In this section, we will talk about each of these elements except color, because color perception is a big subject and deserves a section of it own. Therefore we will talk about color perception in the next section. Here we are concerned with line, shape and form, and texture.

### Designing Visualizations

- How should color be used?

- What graphical entities can be accurately measured?
- How many distinct entities can be used without confusion?

**Computer Vision**

- What primitives do humans detect pre-attentively?
- What level of accuracy do we perceive various primitives?
- How do we combine primitives to recognize complex phenomena?

In any Image Understanding system, there are certain metrics that decide the actual content of the image or the scene. The human distinguishes between 10 graphical perception tasks:

- Angle
- Area
- Color Hue
- Color Saturation
- Density (amount of black)
- Length (distance)
- Position along a common scale
- Position along identical, nonaligned scales
- Slope
- Volume

Weber's Law: likelihood of detection is proportional to the relative change, not the absolute change, of a graphical attribute

Stevens' Law: perceived scale in absolute measurements is the actual scale raised to a power. For linear features power is between .9 and 1.1, for area features it is between .6 and .9, and for volume features it is between .5 and .8.

Experiments showed errors in perception ordered as follows (increasing error)

- Position along a common scale

- Position along identical, non-aligned scales

- Length

- Angle/Slope (though error depends greatly on orientation and type)

- Area

- Volume

- Color Hue, Saturation, Density (only informal testing)

## Form Perception

### Line

A line is the path made by a pointed instrument, such as a pen, a crayon, or a stick. A line implies action because work needs to be done to make it. Moreover, the impression of movement suggests sequence, direction, or force. In other words, a line can be seen as a distinct series of points.

Line is believed to be the most expressive of the visual elements because of several reasons. First, it outlines things and the outlines are key to their identity. Most of the time, we recognize objects or images only from their outlines. Second, line is important because it is a primary element in writing and drawing, and because writing and drawing are universal. Third, unlike texture, shape and form, line is unambiguous. We know exactly when it starts and ends. Finally, line leads our eyes by suggesting direction and movement.

### Shape

Shape is related to line. Closed lines become the boundaries of shapes. The shapes that artists create are inspired by many different sources, such as nature and man-made objects. Like with lines, there are many ways of categorizing shapes. We can use their dimensions, for example, distinguishing between two-dimensional shape and three-dimensional shape. Or we can use their style (realism, abstraction, etc), or their origin (organic or geometric)to classify them.

## Texture

Texture is an element of art that refers to the way things feel, or look as though they might feel, if touched. For example, sandpaper looks and feels rough; a cotton ball looks and feels soft. The connection between visual and tactile sensation is very well developed.

# Color Perception

Our sensations of colour are within us and colour cannot exist unless there is an observer to perceive them. Colour does not exist even in the chain of events between the retinal receptors and the visual cortex, but only when the information is finally interpreted in the consciousness of the observers (Wright, 1963, p. 20).

## Nature of color

What we perceive as color is primarily the wavelength the light stimulation. The shortest viewable wavelength (about 380 nm) is what we see as blue and the longest wavelength (about 760 nm) is what we see as red. The other wavelengths that fall between them are what we see as other colors, as shown in the figure below. However, color perception is very subjective. We do not have a way of proving that two different people perceive the same color, yet we refer to 760-nm wavelength as RED and 380-nm wavelength as BLUE.

## The dimensions of color

Even though wavelength explains differences in the colors we see around us, color entails more than that. There are three psychological dimensions of color: Hue, Brightness, and Saturation. Hue is what we usually refer to as color, therefore, most people use the two words interchangeably. We recognize a change in hue as color change. The physical dimension of hue is wavelength. Brightness is another psychological dimension that refers to the intensity of the stimulus. The more intense the light, the brighter that object appears. For example, an object's color appears brighter in a well-lit room than in a dark one. Saturation is related to the physical dimension of spectral purity. It tells us the amount of hue that we see in an object. In other words, it refers to how complex the light wave is. If the light is simple (for example, a sine wave light), it is pure and therefore appears to be very saturated. The pure color generated by a single wavelength is called monochromatic color.

## Memory color

Even though there is a strong relation between what we perceive as color and the physical characteristics of light stimuli, our perception of color is also influenced by other factors. Examples of these factors are familiarity and past experience. For example, Duncker (1938) found that a green paper cut in a leaf shape is perceived to be greener than the same green paper cut in a donkey shape. This is because leaves are typically green but donkeys are not. Therefore, we can conclude that sometimes previous color and form associations have a strong effect on perceived color.

# Image Processing

## Edge Detection

Edge detection played a major role in visual cognition. This is basically a 3 phase task. The phases are as follows:

- Collecting edge data from human subjects, to obtain ground truths and to adjust the parameters of edge detection methods.

- In phase two the edge detection algorithm is confirmed to be optimized. If not then measures take place to make it so.

- In phase three, all the possible edge detection techniques are applied to the image. The best result is the chosen one.

Edge detection is a problem of fundamental importance in image analysis. In typical images, edges characterize object boundaries and are therefore useful for segmentation, registration, and identification of objects in a scene. In this section, the construction, characteristics, and performance of a number of gradient and zero-crossing edge operators will be presented. An edge basically shows a sudden change in intensity of the image i.e. the pixel values at the edges change suddenly.

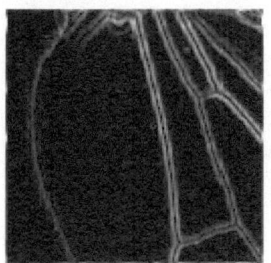
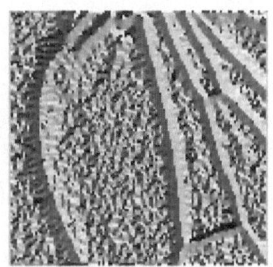

Detected Edges        Original Image

Here is a code snippet of Homogenity Edge Detection Algorithm in C#.

```csharp
public static bool EdgeDetectHomogenity(Bitmap b,
byte nThreshold) {
    Bitmap b2 = (Bitmap) b.Clone();
    BitmapData     bmData     =     b.LockBits(new
Rectangle(0, 0, b.Width, b.Height),
                        ImageLockMode.ReadWrite,
PixelFormat.Format24bppRgb);
    BitmapData     bmData2     =     b2.LockBits(new
Rectangle(0, 0, b.Width, b.Height),
```

```csharp
        ImageLockMode.ReadWrite,
    PixelFormat.Format24bppRgb);
        int stride = bmData.Stride;
        System.IntPtr Scan0 = bmData.Scan0;
        System.IntPtr Scan02 = bmData2.Scan0;
        unsafe {
        byte * p = (byte *)(void *)Scan0;
        byte * p2 = (byte *)(void *)Scan02;
        int nOffset = stride - b.Width*3;
        int nWidth = b.Width * 3;
        int nPixel = 0, nPixelMax = 0;
        p += stride;
        p2 += stride;
        for(int y=1;y<b.Height-1;++y) {
        p += 3;
        p2 += 3;
        for(int x=3; x < nWidth-3; ++x ) {
            nPixelMax = Math.Abs(p2[0] - (p2+stride-
    3)[0]);
            nPixel   =   Math.Abs(p2[0]   -   (p2   +
    stride)[0]);
            if (nPixel>nPixelMax) nPixelMax = nPixel;
            nPixel = Math.Abs(p2[0] - (p2 + stride +
    3)[0]);
            if (nPixel>nPixelMax) nPixelMax = nPixel;
            nPixel   =   Math.Abs(p2[0]   -   (p2   -
    stride)[0]);
            if (nPixel>nPixelMax) nPixelMax = nPixel;
            nPixel   =   Math.Abs(p2[0]   -   (p2   +
    stride)[0]);
            if (nPixel>nPixelMax) nPixelMax = nPixel;
            nPixel = Math.Abs(p2[0] - (p2 - stride -
    3)[0]);
            if (nPixel>nPixelMax) nPixelMax = nPixel;
            nPixel   =   Math.Abs(p2[0]   -   (p2   -
    stride)[0]);
            if (nPixel>nPixelMax) nPixelMax = nPixel;
```

```
            nPixel = Math.Abs(p2[0] - (p2 - stride +
3)[0]);
        if (nPixel>nPixelMax) nPixelMax = nPixel;
        if (nPixelMax < nThreshold) nPixelMax = 0;
        p[0] = (byte) nPixelMax;
        ++ p;
        ++ p2;
    }
    p += 3 + nOffset;
    p2 += 3 + nOffset;
    }
    }
    b.UnlockBits(bmData);
    b2.UnlockBits(bmData2);
    return true;
}
```

## Texture Analysis

Image Texture gives us information about the spatial arrangement of color or intensities in an image or selected region of an image.

The use of image texture can be used as a description for regions into segments. There are two main types of segmentation based on image texture, region based and boundary based. Though image texture is not a perfect measure for segmentation it is used along with other measure, such as color, that helps solve segmenting in image.

- Region Based: Attempts to group or cluster pixels based on texture properties together.

- Boundary Based: Attempts to group or cluster pixels based on edges between pixels that come from different texture properties.

Texture Analysis

## Light and the Environment - Optic Flow Patterns

Changes in the flow of the optic array contain important information about what type of movement is taking place. For example:

i ) Any flow in the optic array means that the perceiver is moving, if there is no flow the perceiver is static.

ii) The flow of the optic array will either be coming from a particular point or moving towards one. The center of that movement indicates the direction in which the perceiver is moving.

If a flow seems to be coming out from a particular point, this means the perceiver is moving towards that point; but if the flow seems to be moving towards that point, then the perceiver is moving away. See above for moving towards an object, below is moving away:

An optical flow pattern of a train from its back

Courtesy: www.simplypsychology.org

This is code snippet for calculating the optical flow of a series of tiff images.

```
static void Simple() {
    string dir = "c:/";
    int N = 8;
    string[] filenames = new string[N];
    for(int i = 0; i < N; i++)
    filenames[i] = dir + "anim."+(i+10)+".tif";
    WAdetector WA = new WAdetector();
    float[,,] f =
WA.WAComputeOpticalFlow(filenames);
    float[,,]     quiver_data = new
float[f.GetLength(0),
    f.GetLength(1),2];     // for the quiver plot
    for(int y = 0; y < quiver_data.GetLength(0);
y++)
    for(int x = 0; x < quiver_data.GetLength(1);
x++) {
        quiver_data[y,x,0] = f[y,x,0]*f[y,x,2];
        quiver_data[y,x,1] = f[y,x,1]*f[y,x,2];

    }
    Bitmap bmp = WAdetector.Quiver(quiver_data);
    int date = DateTime.Now.Year*1000 +
DateTime.Now.Month*100 + DateTime.Now.Day;
    int time = DateTime.Now.Hour*1000 +
DateTime.Now.Minute*100 + DateTime.Now.Second;
    string filename = dir + "simple_flow-" + date
+ "-" + time + ".bmp";
    bmp.Save(filename); // save the quiver plot
}

public float[,,] WAComputeOpticalFlow(string[]
filename){
    Bitmap bmp;
    try{
        bmp = new Bitmap(filename[0]);
```

```
    }
    catch{
        throw new Exception("ReadImage: unable to
read file " + filename[0]);
    }
    int h = bmp.Height;
    int w = bmp.Width;
    int x = (int) Math.Round(0.5*(Math.Log(h,2) +
Math.Log(w,2)));
    int n = x - 5;
    if  (n < 0) n = 0;
    if  (n > 3) n = 3;
    int[]    levels_to_compute = new int[]{n};
    float[][,,] f = WAComputeOpticalFlow(80, 2,
16, 10, levels_to_compute, filename);
    return f[0];
}
```

### The Role of Invariants in Perception

We rarely see a static view of an object or scene. When we move our head and eyes or walk around our environment, things move in and out of our viewing fields. Textures expand as you approach an object and contract as you move away.

There is a pattern or structure available in such texture gradients which provides a source of information about the environment. This flow of texture is INVARIANT, i.e. it always occurs in the same way as we move around our environment and, according to Gibson, is an important direct cue to depth. Two good examples of invariants are texture and linear perspective.

### Affordances

They are cues in the environment, that aid perception. Important cues in the environment include:

- **Optical Array**: The patterns of light that reach the eye from the environment.

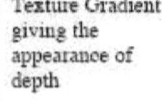

Texture Gradient
giving the
appearance of
depth

Linear Perspective

Parallel lines, eg railway tracks,
appear to converge as they recede
into the distance.

Image Courtesy: www.simplypsychology.org

- **Relative Brightness**: Objects with brighter, clearer images are perceived as closer

- **Texture Gradient**: The grain of texture gets smaller as the object recedes. Gives the impression that surfaces are receding into the distance.

- **Relative Size**: When an object moves further away from the eye the image gets smaller. Objects with smaller images are seen as more distant.

- **Superimposition**: If the image of one object blocks the image of another, the first object is seen as closer.

- **Height in The Visual Field**: Objects further away are generally higher in the visual field

## Segmentation of Images

The goal of image segmentation is to cluster pixels into salient image regions, i.e., regions corresponding to individual surfaces, objects, or natural parts of objects. Segmentation could be used for object recognition, occlusion boundary estimation within motion or stereo systems, image compression, image editing, or image database look-up. We consider bottom-up image segmentation. That is, we ignore (topdown) contributions from object recognition in the segmentation process. For input we primarily consider image brightness here, although similar techniques can be used with colour, motion, and/or stereo disparity information.

The **K-means algorithm** is an iterative technique that is used to partition an image into K clusters. The basic algorithm is:

- Pick K cluster centers, either randomly or based on some heuristic

- Assign each pixel in the image to the cluster that minimizes the distance between the pixel and the cluster center

- Re-compute the cluster centers by averaging all of the pixels in the cluster

- Repeat steps 2 and 3 until convergence is attained (e.g. no pixels change clusters)

In this case, distance is the squared or absolute difference between a pixel and a cluster center. The difference is typically based on pixel color, intensity, texture, and location, or a weighted combination of these factors. K can be selected manually, randomly, or by a heuristic. This algorithm is guaranteed to converge, but it may not return the optimal solution. The quality of the solution depends on the initial set of clusters and the value of K.

Code snippet in C# demonstrating image segmentation using K-Means is shown below:

```
public KMeans(Bitmap bmp, int numCluster,
Colour.Types model)
{
    _image = (Bitmap)bmp.Clone();
    _processedImage = (Bitmap)bmp.Clone();
    _model = model;
```

```
    _previousCluster = new Dictionary<string,
Cluster>();
    _currentCluster = new Dictionary<string,
Cluster>();
    FindTopXColours(numCluster); //find top X
colours in the image
    //create clusters for top X colours
    for (int i = 0; i < _topColours.Length; i++)
    {
        PixelData pd =
        Colour.GetPixelData(_topColours[i].R,
_topColours[i].G, _topColours[i].B, model);

        _previousCluster.Add(_topColours[i].Name,
new Cluster(pd.Ch1, pd.Ch2, pd.Ch3));
        _currentCluster.Add(_topColours[i].Name,
new Cluster(pd.Ch1, pd.Ch2, pd.Ch3));
    }
}

private void CheckConvergence()
{
    //if current and previous cluster centroids
are the same then converged
    bool match = true;
    foreach (KeyValuePair<string, Cluster> cluster
in _currentCluster)
    {
        if (((int)cluster.Value.CentroidR !=
(int)_previousCluster[cluster.Key].CentroidR)
            && ((int)cluster.Value.CentroidG !=
(int)_previousCluster[cluster.Key].CentroidG)
            && ((int)cluster.Value.CentroidB !=
(int)_previousCluster[cluster.Key].CentroidB))
        {
            match = false;
            break;
        }
    }
    if (!match)
    {
        foreach (KeyValuePair<string, Cluster>
cluster in _currentCluster)
        {

_previousCluster[cluster.Key].CentroidR=cluster.Va
lue.CentroidR;

_previousCluster[cluster.Key].CentroidG=cluster.Va
lue.CentroidG;

_previousCluster[cluster.Key].CentroidB =
cluster.Value.CentroidB;
```

```
            }
        }
        _converged = match;
    }

    private void CalculateClusterCentroids()
    {
        foreach (KeyValuePair<string, Cluster> cluster
in _currentCluster)
        {
            List<PixelData> clrList =
(List<PixelData>)_pixelDataClusterAllocation[clust
er.Key];
            float cR = 0;
            float cG = 0;
            float cB = 0;
            foreach (PixelData clr in clrList)
            {
                cR += clr.Ch1;
                cG += clr.Ch2;
                cB += clr.Ch3;

                if
(!_clusterColours.ContainsKey(clr.Name))
                {
                    _clusterColours.Add(clr.Name,
Color.FromArgb((int)cluster.Value.CentroidR,
(int)cluster.Value.CentroidG,
(int)cluster.Value.CentroidB));
                }
            }
            float count = clrList.Count + 1; //total
of colours plus 1 for the existing centroid
            cluster.Value.CentroidR =
(cluster.Value.CentroidR + cR) / count; //average
to find new centroid
            cluster.Value.CentroidG =
(cluster.Value.CentroidG + cG) / count;
            cluster.Value.CentroidB =
(cluster.Value.CentroidB + cB) / count;
        }
    }

    private void AllocateToCluster(PixelData pd)
    {
        //find distance of this colour from each
cluster centroid
        Dictionary<string, Distance> distances = new
Dictionary<string, Distance>();

        foreach (KeyValuePair<string, Cluster> c in
_currentCluster)
        {
```

```
        float d = (float)Math.Sqrt(
            (double)Math.Pow((c.Value.CentroidR -
pd.Ch1), 2) +
            (double)Math.Pow((c.Value.CentroidG -
pd.Ch2), 2) +
            (double)Math.Pow((c.Value.CentroidB -
pd.Ch3), 2)
        );
        distances.Add(c.Key, new Distance(d));
    }

    //allocate this colour to the closest cluster
based on distance
    List<KeyValuePair<string, Distance>> list =
new List<KeyValuePair<string, Distance>>();
    list.AddRange(distances);

    list.Sort(delegate(KeyValuePair<string,
Distance> kvp1, KeyValuePair<string, Distance>
kvp2)
        { return
Comparer<float>.Default.Compare(kvp1.Value.Measure
, kvp2.Value.Measure); });

    //assign to closest cluster
    if
(_pixelDataClusterAllocation.ContainsKey(list[0].K
ey))
    {

((List<PixelData>)_pixelDataClusterAllocation[list
[0].Key]).Add(pd);
    }
    else
    {
        List<PixelData> clrList = new
List<PixelData>();
        clrList.Add(pd);

_pixelDataClusterAllocation.Add(list[0].Key,
clrList);
    }
}

private void FindTopXColours(int numColours)
{
    Dictionary<string, ColourCount> colours = new
Dictionary<string, ColourCount>();
    UnsafeBitmap fastBitmap = new
UnsafeBitmap(_image);
    fastBitmap.LockBitmap();
    Point size = fastBitmap.Size;
    BGRA* pPixel;
```

```
for (int y = 0; y < size.Y; y++)
{
    pPixel = fastBitmap[0, y];
    for (int x = 0; x < size.X; x++)
    {
        //get the bin index for the current
pixel colour
        Color clr = Color.FromArgb(pPixel-
>red, pPixel->green, pPixel->blue);

        if (colours.ContainsKey(clr.Name))
        {
((ColourCount)colours[clr.Name]).Count++;
        }
        else
            colours.Add(clr.Name, new
ColourCount(clr, 1));

        //increment the pointer
        pPixel++;
    }

}
```

## Motion Parallax

As the eyes of humans and other animals are in different positions on the head, they present different views simultaneously. This is the basis of stereopsis, the process by which the brain exploits the parallax due to the different views from the eye to gain depth perception and estimate distances to objects. Animals also use motion parallax, in which the animals (or just the head) move to gain different viewpoints. For example, pigeons move their heads up and down to see depth.

## Object Recognition

### Haar Classifiers

Object detection system is given an image patch of known size or a feature and is to decide whether this features stemmed from an object, or a non object. For the purpose to get a reasonable accuracy of object detection performance, the Haar- classifier is applied to this system

Haar-Classifier encodes the existence of oriented contrasts between regions in the image. A set of these features can be used to encode the contrasts exhibited by an object. The detection technique is based on the idea of the wavelet template that defines the shape of an object in terms of a subset of the wavelet coefficients of the image. Haar-like features are so called because they share an intuitive similarity with the Haar wavelets.

Historically, for the task of object recognition, working with only image intensities (i.e. the RGB pixel values at each and every pixel of image) made the task computationally expensive. This feature set considers rectangular regions of the image and sums up the pixels in this region. The value this obtained is used to categorize images. For example, let us say we have an image database with human faces and buildings. It is possible that if the eye and the hair region of the faces are considered, the sum of the pixels in this region would be quite high for the human faces and arbitrarily high or low for the buildings.

Haar-like Features

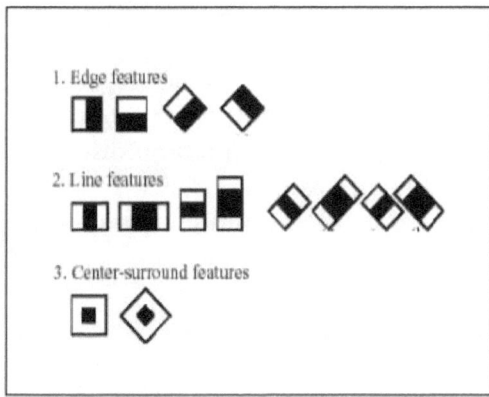

Extended Haar-like Features

## Speeded Up Robust Features

The SURF algorithm is composed of three consecutive steps:
- interest point detection,
- interest point description,
- Feature matching.

Like the SIFT method, the first two steps rely on a scale-space representation and first and second order differential operators. The originality of the SURF method is that these operations are speeded-up by the use of an integral image and box filters techniques detailed in Sections Scale-space approximation and interest point selection respectively.

In the detection step, the local maxima of the Hessian determinant operator applied to the scale-space are computed to select interest point candidates. These candidates are then validated if the response is above a given threshold. Both scale and location of these candidates are then refined using an iterated procedure to fit a quadratic function. Typically, a few hundreds interest points are detected in a digital image of 1 Mega-pixels.

The purpose of the second step described in the local representation section is to build a descriptor that is invariant to view-point changes of the local neighborhood of the point of interest. Recall that the location of this point in the scale-space provides invariance to scale and translation changes. To achieve rotation invariance, a dominant orientation is defined by considering the local gradient orientation distribution, estimated with Haar wavelets. Making use of a spatial localization grid, a 64-dimensional descriptor is then built, corresponding to a local histogram of the Haar wavelet responses.

Classically, the third step matches the descriptors of both images. Exhaustive comparisons are performed here by computing Euclidean distance between all potential matching pairs.

# Reconstruction from Multiple 2d Images

3D reconstruction from multiple images is the creation of three dimensional models from a set of images. It is the reverse process of obtaining 2D images from 3D scenes. The essence of an image is a projection from a 3D scene onto a 2D plane, during which process the depth is lost. The 3D point corresponding to a specific image point is constrained to be on the line of sight. From a single image, it is impossible to determine which point on this line corresponds to the image point. If two images are available, then the position of a 3D point can be found as the intersection of the two projection rays. This process is referred to as triangulation.

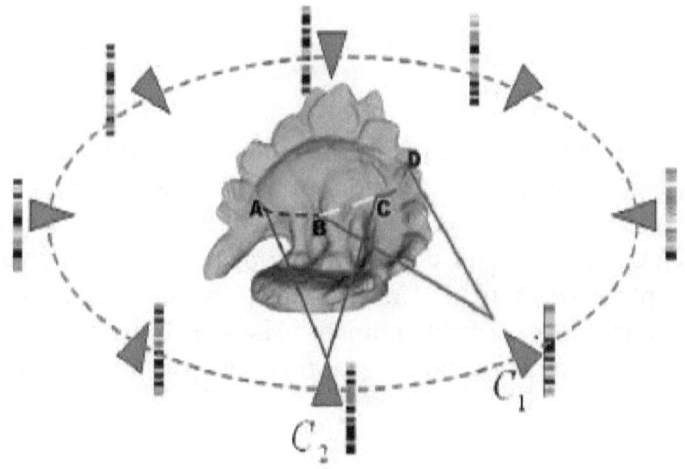

3D Reconstruction using Multiple Images

Image Courtesy: media.au.tsinghua.edu.cn

The key for this process is the relations between multiple views which convey the information that corresponding sets of points must contain some structure and that this structure is related to the poses and the calibration of the camera.

In recent decades, there is an important demand for 3D content for computer graphics, virtual reality and communication, triggering a change in emphasis for the requirements. Many existing systems for constructing 3D models are built around specialized hardware (e.g. stereo rigs) resulting in a high cost, which cannot satisfy the requirement of its new applications. This gap stimulates the use of digital imaging facilities (like a camera). Moore's law also tells us that more work can be done in software. An early method was proposed by Tomasi and Kanade. They used an affine factorization approach to extract 3D from images sequences. However, the assumption of orthographic projection is a significant limitation of this system.

This brings us to the end of computer vision and now we will discuss about robotics and its applications.

# Robotics

Robotics is the branch of technology that deals with the design, construction, operation and application of robots and computer systems for their control, sensory feedback, and information processing. These technologies deal with automated machines that can take the place of humans, in hazardous or manufacturing processes, or simply just resemble humans. Many of today's robots are inspired by nature contributing to the field of bio-inspired robotics.

The concept in creation of machines that could operate autonomously dates back to classical times, but research into the functionality and potential uses of robots did not grow substantially until the 20th century. Throughout history, robotics has been often seen to mimic human behavior, and often manage tasks in a similar fashion. Today, robotics is a rapidly growing field, as technological advances continue; research, design, and building new robots serve various practical purposes, whether domestically, commercially, or militarily. Many robots do jobs that are hazardous to people such as defusing bombs, exploring shipwrecks, and mines.

## Robots

A robot is a mechanical or virtual artificial agent, usually an electro-mechanical machine that is guided by a computer program or electronic circuitry. Robots can be autonomous, semi-autonomous or remotely controlled and range from humanoids such as ASIMO and TOPIO to Nano robots, 'swarm' robots, and industrial robots. By mimicking a lifelike appearance or automating movements, a robot may convey a sense of intelligence or thought of its own. The branch of technology that deals with robots is called robotics.

Machinery was initially used for repetitive functions, such as lifting water and grinding grain. With technological advances more complex machines were developed, such as those invented by Hero of Alexandria in the 1st century AD, and the automata of Al-Jazari in the 12th century AD. The robots made by such inventors were more for the purpose of entertainment than for performing work.

As mechanical techniques developed through the Industrial age, more practical applications were proposed by Nikola Tesla, who in 1898 designed a radio-controlled boat. Electronics evolved into the driving force of development with the advent of the first electronic autonomous robots created by William Grey Walter in Bristol, England in 1948. The first digital and programmable robot was invented by George Devol in 1954 and was named the Unimate. It was sold to General Motors in 1961 where it was used to lift pieces of hot metal from die casting machines at the Inland Fisher Guide Plant in the West Trenton section of Ewing Township, New Jersey.

Robots have replaced humans[citation needed] in the assistance of performing those repetitive and dangerous tasks which humans prefer not to do, or are unable to do due to size limitations, or even those such as in outer space or at the bottom of the sea where humans could not survive the extreme environments. There are concerns about the increasing use of robots and their role in society. Robots are blamed for rising unemployment as they replace workers in some functions. The use of robots in military combat raises ethical concerns. The possibility of robot autonomy and potential repercussions has been addressed in fiction and may be a realistic concern in the future.

## Basic Components

Almost anything that would be considered a robot has the following basic elements:

- A moveable body - Robots may have wheels, limbs connected by mechanical joints, or other types of moveable segments.

- An actuator - In order to be activated, robots may use an electric motor, a hydraulic system, a pneumatic system or a combination of all three.

- A power source - A robot needs a power source to drive its actuators. Electric robots use batteries or extension cord. Hydraulic robots need pumps to pressurize the hydraulic fluid, and pneumatic robots need air compressors.

- An electrical circuit - The electrical circuit powers the electric motor, solenoid or valves that control hydraulic or pneumatic systems.

- A reprogrammable brain (computer) - The computer controls all other components. In order to change the robot's behavior, you just have to reprogram the computer.

- A sensory system - Some robots have the ability to collect information about their environment and react to it.

# Robotic Perception

It deals with the tasks of enabling a robot to perform human like perception such as localization, planning, environment mapping and machine learning.

First let's discuss about behavior based approach. A major tenet of the behavior-based approach is action-oriented perception where the task frames the perceptual input. Perception without the context of action is meaningless. Accordingly, expectations about the task and environment should be used to direct perception, focusing attention on information determined to be crucial for the current task. For example, before the vision system on a robot named "Yippy" can recognize gestures, it first must perform 2-D image comparisons to identify motion and then center in on the locus of that motion - the palm before determining the gesture. No perceptual system should try to take in every aspect of an environment. Humans actually parse very little of the environmental data available to them. Good perception depends much more on 'what' than 'how much.' The learning disability, ADD (Attention Deficit Disorder, is caused by the child's attempts to absorb too much available stimulation. It is the ability to discriminate between important and unimportant environmental characteristics that are the key to both biological and artificial learning.

At times, it is impossible to produce accurate data, even about the environmental information deemed crucial. This brings us to another crucial characteristic of good perceptual schemas — the ability to report uncertainty. If a controller is aware of a perceptual deficit, it can work to redirect perceptual resources accordingly. Controllers should be able to trade off perception schemas based on new needs and changes in the environment. A robot trying to stay on a windy, narrow path may need to focus on the edges of the path, whereas the same robot may need to focus elsewhere if it finds itself on a wide road with many moving vehicles. Perception should be actively generated on a need-to-know basis.

Active perception allows perceptual processes to control themselves. For instance, a vision system should control its own cameras, deciding when to swivel, zoom, focus, etc. On the humanoid robot COG, there are two cameras, one that provides a broad view of the environment and another that focuses in on an area of interest. Perceptual schemas should be able to permit integration of diverse sensory modes such as sonar, infrared, laser scanners, ultrasound, vision, thermal, etc. With this use of multi-modal perception, it has become even more important to focus attention on crucial aspects of the environment. For many systems, a filtering mechanism can actively eliminate or choose certain modes of perception for a particular behavior. Perceptual fission can separate perceptual streams for specialized use by separate behaviors. On the other hand, fusion can combine percepts for a particular behavior. For instance, the SFX architecture uses an investigative phase to predict perceptual needs and strategically reconfigure its sensors, and then a performativity phase to fuse raw sensor data, preprocessing it into a percept for a particular behavior. (Murphy & Arkin 1992)

Using such behavior-based sensory strategies, researchers at Carnegie-Mellon have created agents capable of driving real cars across America at speeds above 100 kph. (Pomerleau 1995) Robotic heads are another area where sensing must involve complex sensory integration. The humanoid robot COG is designed to have modes of perception similar to those of a human. The belief is that only an agent that perceives the world like a human will be better enabled to develop human-like intelligence. Perception influences learning, and if we want a robot to be able to learn from humans – either through emulation or through some other exchange of knowledge and skills – it must be able to relate its view of the universe to ours and map its body to ours. In an attempt to model human perception, researchers at MIT have given COG proprioception – a feeling of where and how body parts are oriented. When COG moves a limb, it receives a wealth of feedback information indicating the success of the motion in regard to the intent. In addition, they have attempted to model the human vestibulo-ocular reflex, which allows the eyes to remain focused on a target even while the head moves.

## Localization

Localization involves one question: Where is the robot now? Or, robo-centrically, where am I, keeping in mind that "here" is relative to some landmark (usually the point of origin or the destination) and that you are never lost if you don't care where you are.

Although a simple question, answering it isn't easy, as the answer is different depending on the characteristics of your robot.

Localization techniques that work fine for one robot in one environment may not work well or at all in another environment. For example, localizations which work well in an outdoors environment may be useless indoors. All localization techniques generally provide two basic pieces of information:

- what is the current location of the robot in some environment?

- what is the robot's current orientation in that same environment?

The first could be in the form of Cartesian or Polar coordinates or geographic latitude and longitude. The latter could be a combination of roll, pitch and yaw or a compass heading.


In order to autonomously navigate and perform useful tasks, a mobile robot needs to know its exact position and orientation. Robot localization is thus a key problem in providing autonomous capabilities to a mobile robot. The different techniques that have been developed to tackle this problem can be classified into two main categories:

- Relative (local) localization: evaluating the position and orientation using information provided by various on-board sensors (e.g. encoders, gyroscopes, accelerometers etc).

- Absolute (global) localization: obtaining the absolute position using beacons, landmarks or satellitebased signals (e.g. GPS).


A popular local technique, dead reckoning, employs simple geometric equations (a kinematic model of the robot) on odometric data to compute the position of the robot relative to its start position. Dead reckoning cannot be used for long distances because it suffers from various drawbacks. The kinematic model always has some inaccuracies, encoders have limited precision and there are external sources affecting the motion that are not observable by the sensors (e.g. wheel slippage). The localization error grows with time.

Applying Kalman filter techniques can provide substantial improvement to the final output of the applied procedure.

## Monte Carlo Localization

Monte Carlo Localization, also known as Particle Filtering, is a relatively new approach to the problem of robot localization - estimating a robot's location in a known environment, given its movements and sensor reading over time. In this project you are to solve the global localization problem, where the robot does not know its starting position but needs to figure out where it is. (This is in contrast to the position tracking problem, where the robot knows it's starting position and just need to accommodate the small errors in its odometry that build up over time.) To make things a bit simpler, you will solve this problem in a one dimensional world. Since the on-board computation abilities of the RCX are limited, we remote control the robot from a base computer. You are given skeleton programs for the robot and base computer, which are described below.

## Problem:

Imagine a long hallway with a set of open doors along one side. The doors are distributed unevenly along it, and the doors are not all of the same width. Your robot can move back and forth along the hallway, and at any time it is either in front of one of the doors or is along a wall segment. The situation might look like this:

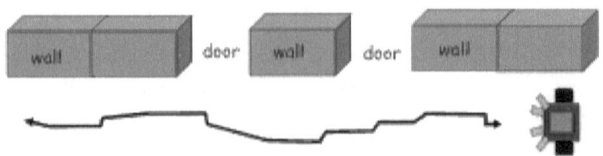

The robot's task is move to a predefined goal point along the hallway. But it doesn't know its starting point. It does have a sonar unit that is aimed at the wall/doors. The unit is reasonably reliable, and can be used to determine whether the robot is currently in front of a wall or a door. To solve this problem, the robot repeated moves forward a fixed distance and takes a sonar reading. If it thinks it has reached the end of the hallway, it starts backing up instead of moving forward. It may need to move back and forth along the hallway a few times before it accurately knows where it is, but usually it will be able to reliably determine its location in one pass back and forth.

Not only does the robot not know its starting position, but there are two additional problems that need to be taken into consideration: The sonar reading may be wrong (e.g., the reading indicates an open door when in front of a wall segment). The distance the robot attempts to move may not the actual distance it moved (e.g. when it attempts to move 2 inches forward it really just moves 1.9 inches). So you will need to develop probabilistic models for both of these sources of error.

## Solution using MCL:

The Monte Carlo approach to localization is based on a collection of samples (which are also known as particles). Each sample consists of a possible location the robot may currently occupy, along with a value which represents the probability that the robot is currently at that location. Strictly speaking these are not probabilities, so they are often referred to as importance weights, and we will use that terminology here.

How many samples should be used? The more you have, the more quickly the program will converge to a correct solution, but at the cost of more computational time and hence slower robot movement. When the program begins the robot does not know where it is, so the current samples (really the locations the samples contain) are evenly distributed over the range of possible locations, and the importance weights are all the same. Over time the samples near the actual current position should become more likely, and those further away less likely. If we created a line graph which plotted the samples using location verses importance weight, the graph should spike over the actual location. Think of this curve as representing the robot's belief state about the robot's location. In the beginning it is a flat line (since the robot has no idea where it is), but over time it should show humps over likely locations, with (hopefully) a large spike over the actual location.

The general idea is as follows:

1.  Initialize the set of samples (the current samples) so that their locations are evenly distributed and their importance weights are equal.

2.  Repeat until done with the current set of samples:

    i.   Move the robot a fixed distance and then take a sensor reading.

    ii.  Update the location of each of the samples (using the movement model).

    iii. Assign the importance weights of each sample to the likelihood of that sensor reading given that new location (using the sensor model).

Remember that the actual distance moved may not be the expected distance. So model that by adding (or subtracting) a random error factor. Similarly, if the sensor reading seems to indicate a door, remember that is might be wrong, and take that into consideration. These are called, respectively, the movement and sensor models. One could stop refining the algorithm here - over time the importance weights would tend towards a solution.

However, the samples with low importance weights add little to figuring out the robot's current location. That fact leads us to the answer of the next question: "Where does the Monte Carlo part come in - doesn't that name suggest games of chance?" Yes, it does. And this is where the big improvement comes in, making the algorithm converge to a solution more quickly. In the loop above, we now add the following steps:

iv.     Create a new collection of samples by sampling (with replacement) from the current set of samples based on their importance weights.

v.      Let this be new collection become the current set of samples.

In step four, samples with a high importance weight are more likely to be chosen than those with low probability, and some sample may be repeatedly chosen. So we (usually) end up with a set of samples with a higher cumulative importance weight that those we had before. This is similar to the approach genetic algorithms take - survival of the fittest!

It is now possible that more than one sample may be at the same location, and clearly samples will cluster around likely locations. So where is the robot likely to be? : At the location (or area) with the most samples.

## Basic MCL Algorithm:

Here is the MCL algorithm (based on the one presented by Dieter Fox in his paper listed below) for generating at time t the next set of samples St+1 from the current set St. The xt are the locations and the wt the probabilities - so an (xt, wt) pair represents a sample. The distance traveled is ut, and the sensor reading is zt. We use superscripts to indicate the individual samples, so xt(i) is the location of sample i at time t. n is the number of samples.

inputs: Distance ut, sensor reading zt., sample set St = { (xt(i), wt(i)) | i = 1,...,n}

for i = 1 to n do          // First update the current set of samples

   xt = updateDist(xt, ut)        // Compute new location

   wt(i) = prob( zt | xt(i) )          // Compute new probability

St+1 = null                          // Then resample to get the next generation of samples

for i = 1 to n do

Sample an index j from the distribution given by the weights in St

Add (xt(j), wt(j)) to St+1        // Add sample j to the set of new samples

return St+1

This is the demonstration of MCL which is one of the most prevalent algorithms implemented in localization.

# Mapping

Robotic mapping is a discipline related to cartography. The goal for an autonomous robot to be able to construct (or use ) a map or floor plan and to localize itself in it. Robotic mapping is that branch of one, which deals with the study and application of ability to construct map or floor plan by the autonomous robot and to localize itself in it.

Evolutionarily shaped blind action may suffice to keep some animals alive. For some insects for example, the environment is not interpreted as a map, and they survive only with a triggered response. A slightly more elaborated navigation strategy dramatically enhances the capabilities of the robot. Cognitive maps enable planning capacities, and use of current perceptions, memorized events, and expected consequences.

**Map representation:**

The internal representation of the map can be "metric" or "topological":

- The metric framework is the most common for humans and considers a two dimensional space in which it places the objects. The objects are placed with precise coordinates. This representation is very useful, but is sensitive to noise and it is difficult to calculate the distances precisely.

- The topological framework only considers places and relations between them. Often, the distances between places are stored. The map is then a graph, in which the nodes corresponds to places and arcs correspond to the paths.

Many techniques use probabilistic representations of the map, in order to handle uncertainty. There are three main methods of map representations, i.e., free space maps, object maps, and composite maps. These employ the notion of a grid, but permit the resolution of the grid to vary so that it can become finer where more accuracy is needed and coarser where the map is uniform.

## Map learning:

Map-learning cannot be separated from the localization process, and a difficulty arises when errors in localization are incorporated into the map. This problem is commonly referred to as Simultaneous localization and mapping (SLAM). An important additional problem is to determine whether the robot is in a part of environment already stored or never visited. One way to solve this problem is by using electric beacons.

## Path planning:

Path planning is an important issue as it allows a robot to get from point A to point B. Path planning algorithms are measured by their computational complexity. The feasibility of real-time motion planning is dependent on the accuracy of the map, on robot localization and on the number of obstacles. Topologically, the problem of path planning is related to the shortest path problem of finding a route between two nodes in a graph.

## Robot navigation

This article describes the state of art navigation systems that exist in the field of mobile Robotics. Outdoor robots can use GPS in a similar way to automotive navigation systems. Alternative systems can be used with floor plan instead of maps for indoor robots, combined with localization wireless hardware. Electric beacons also have been proposed for cheap robot navigational systems.

## Machine learning in perception

Machine learning plays an important role in robot perception. This is particularly the case when the best internal representation is not known. One common approach is to map high-dimensional sensor streams into lower-dimensional spaces using unsupervised machine teaming methods. Such an approach is called low-dimensional embedding. Machine learning makes it possible to learn sensor and motion models from data, while simultaneously discovering a suitable internal representation.

Another machine learning technique enables robots to continuously adapt to broad changes in sensor measurements. Picture yourself walking from a sun-lit space into a dark neon-lit room. Clearly things are darker inside. But the change of light source also affects all the colors: Neon light has a stronger component of green light than sunlight. Yet somehow we seem not to notice the change. If we walk together with people into a neon-Lit room, we don't think that suddenly their faces turned green. Our perception quickly adapts to the new lighting conditions, and our brain ignores the differences.

Adaptive perception techniques enable robots to adjust to such changes. One such example taken from the autonomous driving domain shows an unmanned ground vehicle adapts its classifier of the concept "drivable surface." How does this work? The robot uses a laser to provide classification for a small area right in front of the robot. When this area is found to be fiat in the laser range scan, it is used as a positive training example for the concept "drivable surface."

Methods that make robots collect their own training data (with labels!) are called self-supervised. In this instance, the robot uses machine learning to leverage a short-range sensor that works well for terrain classification into a sensor that can see much farther. That allows the robot to drive faster; slowing down only when the sensor model says there is a change in the terrain that needs to be examined more carefully by the short-range sensors.

# Motion Planning

Motion planning is a term used in robotics for the process of detailing a task into discrete motions. For example, consider navigating a mobile robot inside a building to a distant waypoint. It should execute this task while avoiding walls and not falling down stairs. A motion planning algorithm would take a description of these tasks as input, and produce the speed and turning commands sent to the robot's wheels. Motion planning algorithms might address robots with a larger number of joints (e.g., industrial manipulators), more complex tasks (e.g. manipulation of objects), different constraints (e.g., a car that can only drive forward), and uncertainty (e.g. imperfect models of the environment or robot).

Motion planning has several robotics applications, such as autonomy, automation, and robot design in CAD software, as well as applications in other fields, such as animating digital characters, video game AI, architectural design, robotic surgery, and the study of biological molecules.

## Grid-Based Search

Grid-based approaches overlay a grid on configuration space, and assume each configuration is identified with a grid point. At each grid point, the robot is allowed to move to adjacent grid points as long as the line between them is completely contained within Cfree (this is tested with collision detection). This discretizes the set of actions, and search algorithms (like A*) are used to find a path from the start to the goal.

These approaches require setting a grid resolution. Search is faster with coarser grids, but the algorithm will fail to find paths through narrow portions of Cfree. Furthermore, the number of points on the grid grows exponentially in the configuration space dimension, which makes them inappropriate for high-dimensional problems.

Traditional grid-based approaches produce paths whose heading changes are constrained to multiples of a given base angle, often resulting in suboptimal paths. Any-angle path planning approaches find shorter paths by propagating information along grid edges (to search fast) without constraining their paths to grid edges (to find short paths).

Grid-based approaches often need to search repeatedly, for example, when the knowledge of the robot about the configuration space changes or the configuration space itself changes during path following. Incremental heuristic search algorithms replan fast by using experience with the previous similar path-planning problems to speed up their search for the current one.

## Bug Algorithms

Even a simple planner can present interesting and difficult issues. The Bug algorithms are among the earliest and simplest sensor-based planners with provable guarantees. These algorithms assume the robot is a point operating in the plane with a contact sensor or a zero range sensor to detect obstacles. When the robot has a finite range (non-zero range) sensor, then the Tangent Bug algorithm is a Bug derivative that can use that sensor information to find shorter paths to the goal.

The Bug and Bug-like algorithms are straightforward to implement; moreover, a simple analysis shows that their success is guaranteed, when possible. These algorithms require two behaviors: move on a straight line and follow a boundary. To handle boundary-following, we introduce a curve-tracing technique based on the implicit function theorem at the end of this chapter. This technique is general to following any path, but we focus on following a boundary at a fixed distance. The algorithm is explained below:

**Input:** A point robot with a tactile sensor
**Output:** A path to $q_{goal}$ or a conclusion no such path exists

1: **while** True **do**
2:    **repeat**
3:      From $q_{i-1}^{L}$, move toward $q_{goal}$ along $m$-line.
4:    **until**
     $q_{goal}$ is reached **or**
     an obstacle is encountered at *hit point* $q_{i}^{H}$.
5:    Turn left (or right).
6:    **repeat**
7:      Follow boundary
8:    **until**
9:      $q_{goal}$ is reached **or**
10:      $q_{i}^{H}$ is re-encountered **or**
11:      $m$-line is re-encountered at a point $m$ such that
12:        $m \neq q_{i}^{H}$ (robot did not reach the hit point),
13:        $d(m, q_{goal}) < d(m, q_{i}^{H})$ (robot is closer), and
14:        if robot moves toward goal, it would not hit the obstacle
15:    **if** Goal is reached **then**
16:      Exit.
17:    **end if**
18:    **if** $q_{i}^{H}$ is re-encountered **then**
19:      Conclude goal is unreachable
20:    **end if**
21:    Let $q_{i+1}^{L} = m$
22:    Increment $i$
23: **end while**

## PID Algorithm

A proportional–integral–derivative controller (PID controller) is a generic control loop feedback mechanism (controller) widely used in industrial control systems – a PID is the most commonly used feedback controller. A PID controller calculates an "error" value as the difference between a measured process variable and a desired set point. The controller attempts to minimize the error by adjusting the process control inputs. It is one of the most important algorithm used in efficient robot motion.

The PID controller calculation (algorithm) involves three separate constant parameters, and is accordingly sometimes called three-term control: the proportional, the integral and derivative values, denoted

P, I, and D. Heuristically, these values can be interpreted in terms of time: P depends on the present error, I on the accumulation of past errors, and D is a prediction of future errors, based on current rate of change. The weighted sum of these three actions is used to adjust the process via a control element such as the position of a control valve, or the power supplied to a heating element.

In the absence of knowledge of the underlying process, a PID controller has historically been considered to be the best controller. By tuning the three parameters in the PID controller algorithm, the controller can provide control action designed for specific process requirements. The response of the controller can be described in terms of the responsiveness of the controller to an error, the degree to which the controller overshoots the set point and the degree of system oscillation. Note that the use of the PID algorithm for control does not guarantee optimal control of the system or system stability.

Some applications may require using only one or two actions to provide the appropriate system control. This is achieved by setting the other parameters to zero. A PID controller will be called a PI, PD, P or I controller in the absence of the respective control actions. PI controllers are fairly common, since derivative action is sensitive to measurement noise, whereas the absence of an integral term may prevent the system from reaching its target value due to the control action.

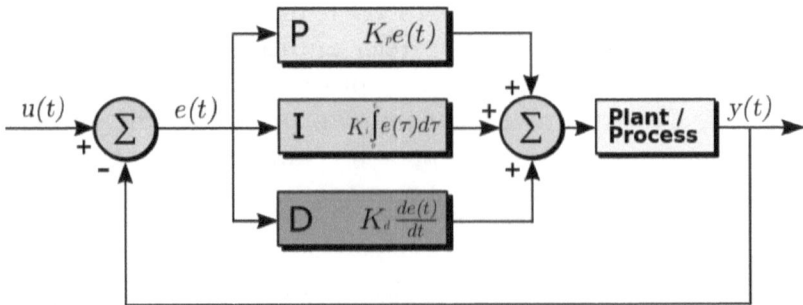

The pseudo code of the algorithm is described below:

Variables:

| | |
|---|---|
| Input | *Process input* |
| InputD | *Process input plus derivative* |
| InputLast | *Process input from last pass, used in deriv. calc.* |
| Err | *Error, Difference between input and set point* |
| SetP | *Set point* |
| OutPutTemp | *Temporary value of output* |
| OutP | *Output of PID algorithm* |
| Feedback | *Result of lag in positive feedback loop.* |
| Mode | *value is 'AUTO' if loop is in automatic* |
| Action | *value is 'DIRECT' if loop is direct acting* |

The PID emulation code:

```
1.   IF Mode = 'AUTO' THEN
2.       InputD=Input+(Input-InputLast)*Derivative*60   derivative.
3.       InputLast = Input
4.       Err=InputD-SetP          Error based on reverse action.
5.       IF Action = 'DIRECT' THEN Err=0 - Err  Change sign if direct.
6.   ENDIF
7.       OutPutTemp = Err*Gain+Feedback Calculate the gain time the error and add the feedback.
8.       IF OutPutTemp > 100 THEN OutPutTemp =100  Limit output to between
9.       IF OutPutTemp < 0 THEN OutPutTemp =0      0 and 100 percent.
10.      OutP = OutPutTemp                The final output of the controller.
11.      Feedback=Feedback+(OutP-Feedback)*ResetRate/60
12.  ELSE
13.      InputLast=Input    While loop in manual, stay ready for bumpless switch to Auto.
14.      Feedback=OutP
15.  ENDIF
```

The complete working of PID algorithm is now demonstrated using a c# code:

```csharp
namespace PIDLibrary
{
    public delegate double GetDouble();
    public delegate void SetDouble(double value);
}
public class PID
{
    #region Fields
    //Gains
    private double kp;
    private double ki;
```

```csharp
private double kd;
//Running Values
private DateTime lastUpdate;
private double lastPV;
private double errSum;
//Reading/Writing Values
private GetDouble readPV;
private GetDouble readSP;
private SetDouble writeOV;
//Max/Min Calculation
private double pvMax;
private double pvMin;
private double outMax;
private double outMin;
//Threading and Timing
private double computeHz = 1.0f;
private Thread runThread;
#endregion
#region Properties
public double PGain
{
    get { return kp; }
    set { kp = value; }
}
public double IGain
{
    get { return ki; }
    set { ki = value; }
}
public double DGain
{
    get { return kd; }
    set { kd = value; }
}
public double PVMin
```

```
    {
        get { return pvMin; }
        set { pvMin = value; }
    }
    public double PVMax
    {
        get { return pvMax; }
        set { pvMax = value; }
    }
    public double OutMin
    {
        get { return outMin; }
        set { outMin = value; }
    }
    public double OutMax
    {
        get { return outMax; }
        set { outMax = value; }
    }
    public bool PIDOK
    {
        get { return runThread != null; }
    }
    #endregion
    #region Construction / Deconstruction
    public PID(double pG, double iG, double dG,
        double pMax, double pMin, double oMax,
double oMin,
        GetDouble pvFunc, GetDouble spFunc,
SetDouble outFunc)
    {
        kp = pG;
        ki = iG;
        kd = dG;
        pvMax = pMax;
        pvMin = pMin;
```

```csharp
        outMax = oMax;
        outMin = oMin;
        readPV = pvFunc;
        readSP = spFunc;
        writeOV = outFunc;
    }


    ~PID()
    {
        Disable();
        readPV = null;
        readSP = null;
        writeOV = null;
    }


    #endregion

    #region Public Methods

    public void Enable()
    {
        if (runThread != null)
            return;

        Reset();

        runThread = new Thread(new
ThreadStart(Run));
        runThread.IsBackground = true;
        runThread.Name = "PID Processor";
        runThread.Start();
    }

    public void Disable()
    {
        if (runThread == null)
```

```
        return;

      runThread.Abort();
      runThread = null;
    }
    public void Reset()
    {
      errSum = 0.0f;
      lastUpdate = DateTime.Now;
    }


    #endregion

    #region Private Methods

    private double ScaleValue(double value, double
valuemin,
            double valuemax, double scalemin,
double scalemax)
    {
      double vPerc = (value - valuemin) /
(valuemax - valuemin);
      double bigSpan = vPerc * (scalemax -
scalemin);

      double retVal = scalemin + bigSpan;

      return retVal;
    }
  private double Clamp(double value, double min,
double max)
    {
      if (value > max)
        return max;
      if (value < min)
        return min;
      return value;
```

```csharp
    }
    private void Compute()
    {
        if (readPV == null || readSP == null ||
writeOV == null)
            return;
        double pv = readPV();
        double sp = readSP();
        //We need to scale the pv to +/- 100%, but
first clamp it
        pv = Clamp(pv, pvMin, pvMax);
        pv = ScaleValue(pv, pvMin, pvMax, -1.0f,
1.0f);
        //We also need to scale the setpoint
        sp = Clamp(sp, pvMin, pvMax);
        sp = ScaleValue(sp, pvMin, pvMax, -1.0f,
1.0f);
        //Now the error is in percent...
        double err = sp - pv;
        double pTerm = err * kp;
        double iTerm = 0.0f;
        double dTerm = 0.0f;
        double partialSum = 0.0f;
        DateTime nowTime = DateTime.Now;
        if (lastUpdate != null)
        {
            double dT = (nowTime -
lastUpdate).TotalSeconds;

            if (pv >= pvMin && pv <= pvMax)
            {
                partialSum = errSum + dT * err;
                iTerm = ki * partialSum;
            }
            if (dT != 0.0f)
                dTerm = kd * (pv - lastPV) / dT;
        }
```

```
        lastUpdate = nowTime;
        errSum = partialSum;
        lastPV = pv;


        //Now we have to scale the output value to
match the requested scale
        double outReal = pTerm + iTerm + dTerm;
        outReal = Clamp(outReal, -1.0f, 1.0f);


        outReal = ScaleValue(outReal, -1.0f, 1.0f,
outMin, outMax);
        //Write it out to the world
        writeOV(outReal);
    }
    #endregion
    #region Threading
    private void Run()
    {
        while (true)
        {
            try
            {
                int sleepTime = (int)(1000 /
computeHz);
                Thread.Sleep(sleepTime);
                Compute();
            }
            catch (Exception e)
            {
            }
        }
    }

    #endregion
}
```

# Application of Robotics

## Industrial Robot

Many industrial robots are available which consist of an anthropomorphic arm (i.e. with a shoulder, elbow and wrist, and appropriate tool or gripper where a hand would normally be), and for initial feasibility trials a Puma robot was indeed tried in the laboratory. However, this is not practical for medical purposes as robot manufactures usually forbid the use of their robots in close proximity to humans (let alone inside humans!). This is because the working environment for such robots is usually large and unconstrained. To circumvent this problem we designed our own robot with a small working environment, constrained to the size of the prostate. This consists of three axes of movement and a fourth axis to move a resectoscope cutter, as shown in the following simplified drawing.

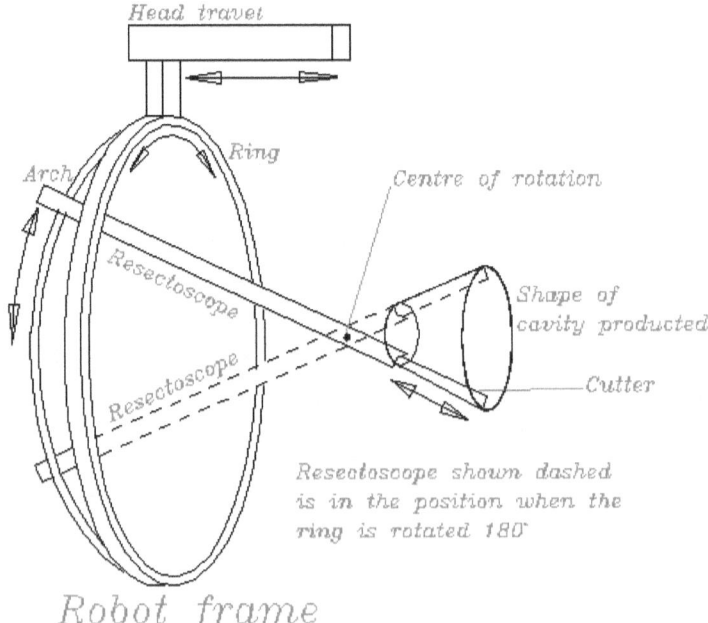

Reselectoscope shown dashed
is in the position when the
ring is rotated 180°

The geometry of such a robot allows a cavity to be hollowed out from within the prostate. The robot is controlled by a pair of programmable embedded motor control systems, which are in turn controlled by a i486DX2 based PC. This provides the user interface and program to generate the correct cutting sequence to remove tissue from the prostate.

# Educational Robots

### Possible roles for robots in education

With robots and related automated process having increasing role in industry they are becoming an object for study in their own right in technology and engineering courses at secondary school and university level. However in this paper we mainly consider the wider role of robots more generally in key elements of the learning process. Robots are a great aid to the teaching of especially maths and physics because of their power to capture the imagination of many younger people. Thus they can be employed to elucidate often difficult abstract concepts. With the robot as the focus of the discussion of a wide range of topics can be brought to life: Newtonian mechanics; measurement; task planning; programming; mathematical formulation of a problem; optimization; limits; etc.

Giving something physical in the 3-dimensional "real world" can help many students grasp the fundamentals of a topic more quickly than just using paper/white board and pen. The robot as well as assisting in conceptualization of a problem provides an environment for experimentation. Possible solutions can be programmed into the robot and then its behavior observed to see if it conforms to that which the student expected. There is then opportunity for iteration towards a correct solution to a particular problem. Thus the power of discovery in effective learning can be readily facilitated through the use of a robot as a teaching aid.

Simple robotic vehicles have been successfully used at both school and university level by both the departments represented by the authors of this paper. At the Open University a simple robotic buggy that can be controlled over the World-Wide-Web has been configured to mimic the NASA Mars Buggy and used extensively with school groups. This was an extension of a remotely controlled robot in a maze that was developed specifically in response to the control technology elements of the English/Welsh National Curriculum for 11-12 year olds [Whalley 1992]. Similarly at the University of Reading autonomous robot programmable "insects" have been used at both school and university level and are more fully described under Experiences with Robot Insects in Education below.

Robots have traditionally been programmed by complex high or even low level computer languages, which would tend to mitigate against their use within education. There are several robot programming languages that have been developed specifically for educational use (e.g. *LOGO as used with the Lego Dacta system). However even these may by overly complex and constrained by the need for precise syntax for many younger or less able students. An interesting approach to address this has been taken and international group of researchers working in evolutionary robotic design. They have successfully demonstrated the use of evolutionary robotic approaches to enable children to design for themselves a range of simple robotic behavior such as collision avoidance, line or wall following, etc. for Lego based mobile robots [Lund, et. al. (1998)]. The same problem has been addressed differently in the work at the Open University. In brief their approach is based on the use of a simple objected orientated control language integrated with a role play and storyboard techniques to enable the children in the creation of their own programmers.

Robots can be an expensive technology with costs ranging from about 100 ECU to 10,000 ECU. Since the use of robots in education is still in its infancy there are difficulties with staff training, technology reliability and a lack of quantitative studies showing the educational impact. Most work reports anecdotally that there is an educational benefit, but there is usually no reliable measure of what factors are causing the benefit. However the authors see robotics as an increasingly available and affordable technology that can address needs of teachers and learners in established areas of the curriculum. They are not attempting to support a technology push approach to hi-tech learning environments.

A survey of any set curriculum for education from the ages of 8 upwards readily yields key opportunities for the application of robotics to those with experience of the pedagogic advantages of approaches based on these technologies.

## Work of Seymour Papert

Seymour Papert, a founding father of this field, supports an approach to learning in the classroom which he calls 'constructionism', opposed to the traditional style of 'instructionism' [Papert, 1993]. By this he means that children will do best by finding or 'fishing' for knowledge by themselves. Improvisational, self-directed, 'playful' activities should simulate the more 'natural' way in which children seem to learn outside the classroom. Instead of a one-way and top-down transmission of knowledge from teacher to child (the behaviorist/objectivist approach), appropriate learning environments ('contexts'), could be used as 'personal media'. This could, according to Papert, empower the child to develop a different relationship to knowledge in a new style of learning, which can account for personal variation in learning styles. In the mid-1960s Papert developed at the MIT AI-Lab with his colleagues the programming language LOGO, a computer language especially designed for children. This is now widely used in control and robotic activities in the classroom. He also went on to develop a programmable computer-sketching device, called a 'Turtle' to introduce mathematical concepts of geometry and shape. Again this has become a widespread technology.

Particular roles for robots with disabled students Educational applications for robots hold particular promise for students or pupils with disabilities in two main ways:

- The robots can be enabling in themselves – students being facilitated to undertake a wide range of tasks that would be otherwise denied them because of their disabilities

- Accessible interfaces to educational robots can lead to disabled students having equal participation with peers in robot based leaning activities

The potential for robots facilitating learning by experiment has already been stated. This approach has added value for the disabled student who may be reduced to an observer role in many conventional student experiments. Provided the appropriate computer interface is available most disabled students can initiate the experiments themselves. Robot Manipulators in Special Education A fully integrated, robot aided, science education programme for students with disabilities was developed by Howell [Howell, et. al. (1994)]. This work was based on a commercial robot (the RTX from OxIM, UK) and focused on developing teaching material based on the US science curriculum for Junior High School students. Examples of teaching material included experiments in biology, where seeds were grown under different conditions, and physics where properties of materials where tested. Harwin and Gosine [Harwin et. al. (1986), Gosine et. al. (1990)] carried out similar work, which had a greater focus on the interface between the person and the robot.

This system was also based on the RTX robot and children with special needs were evaluated in a structured teaching environment. Tasks undertaken in this system ranged from illustrating basic concepts such as block play, problem solving and sequencing tasks, through to simple chemistry experiments, and making and eating simple desserts. An observation of this system is that once the individual was familiar with the interface, they were prepared to experiment, both with the robot and with the environment. One example is that when time for free play was allowed, one student experimented with pouring water from one container to another discovering how water flowed and getting splashed in the process - this proved to be a powerful learning experience.

This illustrates a strong advantage of a robot-based system compared to a software simulation in that the real world has many more interacting factors that cannot be illustrated by a computer. Further it is a demonstration of the robot equaling access for students with special needs to the same equipment used by their peers. A programme of work has recently been begun, led by the Open University, towards developing a flexible learning environment based on remote controlled experimentation. Key objectives for this work are to develop an experimental facility that will enable the active participation of disabled students in science education alongside their peers and to provide a facility that supports the practical elements of science education at a distance. An important feature of this work is that from a standard personal computer students are able to design and configure experiments that they then conducted at a remote laboratory. Robot manipulators and other related technologies have a key role here. As well as the approach of linking robot assisted exercises to a formal syllabus the advantages of a robot being available for free play or exploration should be noted. Students with severe physical disabilities may have missed much from such experiences in their preschool lives because of their inability to interact with their environment and the objects within it in a controlled fashion (e.g. playing in the bath). Thus, robots can have a key role in replacing the informal learning received by most children as they play.

## Mobile Robots in Autism

One of the authors is studying how to use interactive, mobile robots as therapeutic devices for children who have difficulty in co-ordinated interactions with the environment and other people [Dautenhahn 1999]. The project Aurora (Autonomous robotic platform as a remedial tool for children with autism) is using a commercially available mobile robotic platform. The platform itself is seen as a mediator device, i.e. it is intended to encourage children to interact with the environment. Basic forms of social interaction like attraction and avoidance are elements in the robot's interaction repertoire.

## Experiences with Robot Insects in Education

In the Department of Cybernetics at the University of Reading, a set of small robots known as the seven dwarves have been developed specifically for educational use. These robots have been used extensively in education from the ages of 6 through to post-doctoral research over a period of about 6 years. The "dwarves" consist of a 3-wheeled machine, roughly 150mm x 150mm x 150mm driven by 2 electric motors, which can be controlled by the resident software. They also include a set of ultra-sonic sonar sensors by which they sense the world. The way the "dwarves" behave given the sensor information can be fully programmed by the students, at different levels as applicable to the age group. When programmed, the "dwarves" are then set free and their behaviour observed. A lot of fun and educational benefit has resulted from groups of these being programmed and set free to run together. (E.g. groups of student can work together to programme a pair of robots which will follow each other).

The robots thus present an achievable challenge to a wide age range. They are perceived as fun as they travel at speeds up to 1m/sec and can sense obstacles at a distance. The students find that deliberately programming the robots to crash is boring as they only move the once, however programming them to just miss is more challenging and far more enjoyable. Once the students have started to think in a reactive programming way they seem to find it far more intuitive than, for example, programming a sequence of Cartesian commands. The slow response time of the robot of up to 1/10 of a second compared to its speed of up to 1m/sec means the robots' actions are far less deliberate than the students expect and they tend to associate this with character. A robot with a following programme is often described as curious or frightened and often likened to a puppy.

This is an example of a robot-based approach suitable for a wide range of educational circumstances, which can be readily made available to many disabled students by simply providing the appropriate interface to the computer used to programme the robot insects.

# Robotics in Agriculture

## Seed bed preparation

Ploughing is one of the most important primary cultivation processes and has been carried out since the start of civilization. It is effectively the inversion or mixing of topsoil to prepare a suitable seed bed. It also has the ability to bury surface crop residues and control weeds. A small robot utilising current technology does not have the energy density to sustain ploughing over a large area due to the high levels of energy needed to cut and invert the dense soil. Secondly, the draft force required to plough also needs relatively high weight to give traction. Perhaps we would leave it at that, but by considering what the plant, or in this case the seed actually needs, we can approach the problem in a different way. The seed requires contact with the soil moisture to allow uptake of water and nutrients, it requires stability to hold the growing plant and a structure that allows the roots to develop and the shoots to grow. A solution is a two-fold process.

Firstly if we do not compact the soil in the first place there is less need for energy inputs for remedial loosening. Natural soil flora and fauna can be encouraged to manipulate the soil to give a good structure. This is one of the reasons to opt for smaller machines.

Secondly, if the majority of the soil rooting depth is acceptable, then only the local environment of the seed needs to be conditioned before seed placement, which will take a lot less power. Add to this the ability to place nutrients in the correct proximity to the seed we can improve the early phase of establishment. This system has many of the advantages of direct drilling but incorporation of previous crop residues may still cause a problem although removal of crop residues is an option.

## Seed mapping

Seed mapping is the concept of passively recording the geospatial position of each seed as it goes into the ground. It is relatively simple in practice as an RTK GPS is fitted to the seeder and infra-red sensors mounted below the seed chute. As the seed drops, it cuts the infrared beam and triggers a data logger that records the position and orientation of the seeder. A simple kinematic model can then calculate the actual seed position (Griepentrog et al. 2003). The seed coordinates can then be used to target subsequent plant based operations.

## Seed placement

Rather than just record the position of each seed it would be better to be able to control the seed position. This would allow not only allow the spatial variance of seed density to be changed but also have the ability to alter the seeding pattern. Most seeds are dropped at high densities within each row, whilst having relatively more space between the rows. From first agronomic principles, each plant should have equal access to spatial resources of air, light, ground moisture, etc. Perhaps a hexagonal or triangular seeding pattern might be more efficient in this context. If suitable controls are fitted to allow synchronisation between passes, then there is the possibility to plant seeds on a regular grid that can allow orthogonal inter-row weeding. Tests of such a machine will be carried out at KVL in 2005.

## Reseeding

Reseeding is the concept of being able to identify where a seed was not planted, or that a crop plant has not emerged and a machine can automatically place another seed in the same position. This concept could be extended to transplanting a seedling instead of a seed if the surrounding plants are too far advanced. A reseeder would have the ability to insert individual seeds/plants without disturbing the surrounding crop. Conventional seeders could not then be used as they create continuous slots in the soil. A punch planter could be developed to fulfill this role, or better still adapt a Japanese trans-planter to deal with one seedling at a time. Prior local micro-cultivation could be achieved by using a targeted water jet (or gel) to pierce the soil and soften it ready for the seedling roots.

## Crop scouting

One of the main operations within good management is the ability to collect timely and accurate information. Quantified data has tended to be expensive and sampling costs can quickly outweigh the benefits of spatially variable management. (Godwin et al. 2001) Data collection would be less expensive and timelier if an automated system could remain in the crop carrying a range of sensors to assess crop health and status. A high clearance platform is needed to carry instruments above the crop canopy and utilize GPS. Smaller sub canopy machines have been developed in student competitions.

Courtesy (www.fieldrobot.nl)

A range of sensors have been fitted to measure crop nutrient status and stress (multi spectral response), visible images (pan chromatic), weed species and weed density.

## Weed mapping

Weed mapping is process of recording the position and preferably the density (biomass) of different weed species using aspects of machine vision. One method is to just record the increased leaf area found in weedy areas as weeds are patchy and the crops are planted in rows (Pedersen 2001). Another more accurate method is to use active shape recognition, originally developed to recognise human faces, to classify weed species by the shape of their outline (Søgaard and Heisel 2002). Current research has shown that up to 19 species can be recognised in this way. Color segmentation has also shown to be useful in weed recognition (Tang et al. 2000). The final result is a weed map that can be further interpreted into a treatment map.

## Robotic weeding

Knowing the position and severity of the weeds there are many methods that can kill, remove or retard these unwanted plants (Nørremark and Griepentrog 2004) Different physical methods can be used that rely on physical interaction with the weeds. A classic example is to break the soil and root interface by tillage and promote wilting of the weed plants. This can be achieved in the inter row area easily by using classical spring or duck foot tines. Intra row weeding is more difficult as it requires the position of the crop plant to be known so that the end effector can be steered away. Within the close-to-crop area, tillage cannot be used as any disturbance to the soil is likely to damage the interface between the crop and the soil. Non-contact methods are being developed such as laser treatments (Heisel 2001) and micro-spraying. Controlled biodiversity is an opportunity that could be realized with robotic weeding. Non-competitive weeds can be left to grow when they are at a distance from the crop. This is part of the design parameters for the Autonomous Christmas Tree weeder being developed at KVL laboratory.

# Robotics in Medicine

## Biological Applications

The primary purpose for use of robotics in biology is to achieve high throughput in experiments related to research and development of life science. Those experiments involve the delivery and dispensation of biological samples/solutions in large numbers each with very small volumes. Typical applications include high-throughput systems for large-scale DNA sequencing, single nucleotide polymorphism (SNP) analysis, haplotype mapping, compound screening for drug development, and bio-solution mixing and dispensing for membrane protein crystallization. Without robots and automation, biosamples/solutions must be handled manually by human hands, which is not only tedious but also slow. Various robotic systems have been developed in laboratories that are either specially developed for a particular application. The second purpose of robotics for biological applications is for effective handling and exploration of molecular and cell biology. This type of application includes immobilization of individual cells, cell manipulation, and cell injection for pronuclei DNA insertion. Special tools fabricated using different technologies have to be developed such as lasers for micro-sensing and manipulating, electro-active polymer for cell manipulation, and micro-needles for cell penetration. Another interesting area of application is robotics-inspired algorithms for molecular and cellular biology. This includes the work for predicting protein folding, and for structural biology (Zheng and Chen, 2004).

## Medical Applications

Research on robotics for medical applications started fifteen years ago and is very active today. The purpose is three-fold. First it is for robotic surgery. Robotic surgery can accomplish what doctors cannot because of precision and repeatability of robotic systems. Besides, robots are able to operate in a contained space inside the human body. All these make robots especially suitable for non-invasive or minimally invasive surgery and for better outcomes of surgery.

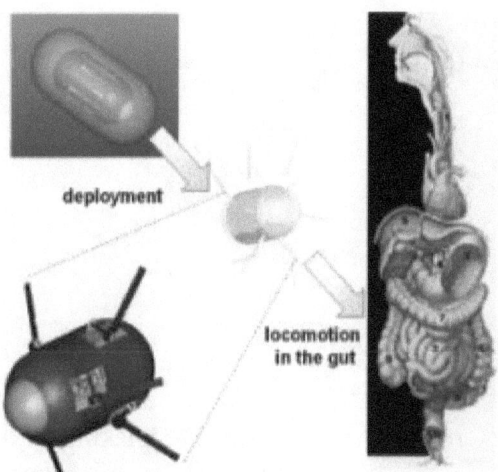

Robotic capsular endoscope for examination of gastrointestinal tract

Today, robots have been demonstrated or routinely used for heart, brain, spinal cord, throat, and knee surgeries at many hospitals in the United States (International Journal of Emerging Medical Technologies, 2005). Since robotic surgery improves consistency and quality, it is becoming more and more popular.

The second use of robotics in medicine is diagnosis. Robotic diagnosis reduces invasiveness to the human body and improves the accuracy and scope of the diagnosis. One example is the robotic capsular endoscope that has been developed for non-invasive diagnosis of gastrointestinal tract by Polo Sant'Anna Valdera of the Sant'Anna School of Advanced Studies in Italy.

## Robotic Tools, Devices and Systems

Robotics for biological and medical applications uses many tools, devices, and systems of both general purpose and specially designed types.

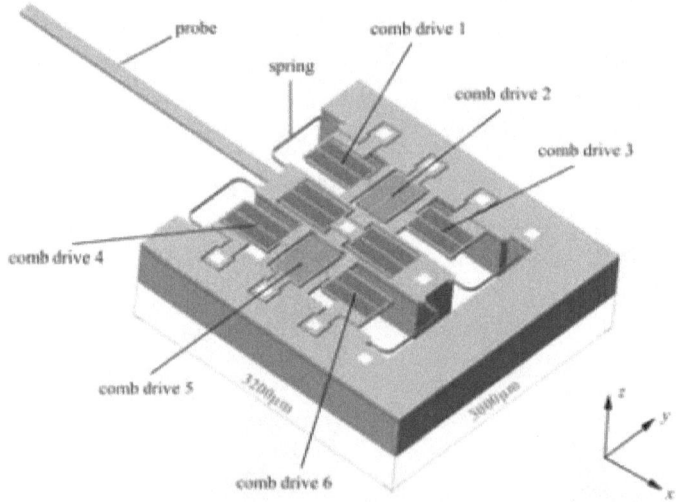

Microforce sensor using integrated circuit (IC) fabrication technology, U. of Minnesota (Nelson and Zheng, 2004).

The former includes robot manipulators for picking and placing, and micro-actuators for dispensing biosamples/solutions. Another example is the system developed by the Novartis Research Foundation's Genomics Institute, which includes standard industrial manipulators for high-throughput screening of compounds up to 1 million samples per day (Meldrum and Kavraki, 2004). In robotic surgery, commercially available robots are often a part of an integrated system. Special purpose sensors have even more types including visual, force, and neuro-sensing. Biosensors often are very small and so Micro-electromechanical systems (MEMS) technology is used to fabricate such elements as the micro-force sensor from the University of Minnesota and ETH-Zürich shown in above figure. Special tools using nontraditional principles are also developed to handle bio-solutions or to manipulate cells. For example, Nagoya University in Japan used local photo polymerization on a chip to immobilize individual cells. Besides tools and devices, software and algorithms are also an important part of robotics for biological and medical applications. In robotic surgery, for example, effective algorithms for modeling and analysis of human body components are an important topic of research. The purpose is to develop patient-specific models for performing precise surgery.

## Key Technologies

Key technologies for robotics in biological and medical applications include the following:

a) MEMS technologies that can fabricate tools and devices suitable for micro-sensing, micro-actuation and micromanipulation of biosamples/solutions and bio-objects such as cells. These technologies use either IC-fabricating methods or use micromachining methods.

b) Special robotic systems that can perform surgery precisely and at low cost. The challenge is to program motion of robots efficiently based on patient-specific modeling and analysis.

c) Modeling and analysis algorithms that is precise and fast for individual patients.

d) Reliable and efficient system integration of off-the-shelf components and devices for specific biological and medical operations.

e) Engineering modeling of biological systems. The purpose is to develop mathematical models for explaining the behavior and structure of biological systems as engineers do for artificial physical systems. This has been proved extremely challenging because of the complexity of biological systems.

f) Solid understanding of life science. To develop an effective robotic or automation system for biological and medical applications, it is necessary for engineers to have a deep understanding of life science.

From the above, one can see that robotics for biological and medical applications cover a wide scope of technologies from conventional robots and sensors to micro sensors and actuators, from tools and devices to algorithms. For molecular-level study of biological systems, nano-devices and actuation are key technologies as well.

# Gaming Applications

Game artificial intelligence refers to techniques used in computer and video games to produce the illusion of intelligence in the behavior of non-player characters (NPCs). The techniques used typically draw upon existing methods from the field of artificial intelligence (AI). However, the term game AI is often used to refer to a broad set of algorithms that also include techniques from control theory, robotics, computer graphics and computer science in general.

Since game AI is centered on appearance of intelligence and good gameplay, its approach is very different from that of traditional AI; workarounds and cheats are acceptable and, in many cases, the computer abilities must be toned down to give human players a sense of fairness. This, for example, is true in first-person shooter games, where NPCs' otherwise perfect aiming would be beyond human skill. Some examples are briefed below:

## Computer Chess

Computer chess is computer architecture encompassing hardware and software capable of playing chess autonomously without human guidance. Computer chess acts as solo entertainment (allowing players to practice and to better themselves when no sufficiently strong human opponents are available), as aids to chess analysis, for computer chess competitions, and as research to provide insights into human cognition. Current chess engines are able to defeat even the strongest human players under normal conditions. Whether computation could ever solve chess remains an open question.

## Call of Duty

Call of Duty: Modern Warfare 3 is a first-person shooter video game, developed by Infinity Ward and Sledgehammer Games (Treyarch for the Wii version), with Raven Software having assisted in development. It is the third installment in the Modern Warfare series, a direct sequel to 2009's Call of Duty: Modern Warfare 2, and the eighth Call of Duty installment. It is one of the latest and most developed game applications which use almost all of the AI algorithms and its latest updated intelligent systems.

# Further Scope

We have written this book to demonstrate the process adapted by intelligent agents and system to imitate human as far as possible. Maybe, in future these systems might overcome the intelligence and emotional level of human and become something more. But to achieve that, we have to keep ourselves update and well educated about the past, latest and possibly incoming concepts in the field of artificial intelligence and some other related areas. We will like to throw light on some more concepts and topics (not covered in this edition) in the later editions of this book series.

# Bibliography

[1]     Cherry, Kendra. "Memory - Overview of Memory." *About.com*. N.p.. Web. 15 July 2012.

[2]     Problem Solving." *Cognitive Psychology*. Wikidot.com, 14 4 2009. Web. 21 Aug 2012.

[3]     Harpaz, Yehuda. "Human Cognition in the Human brain.." . Maldoo.com, 24 10 1999. Web. 25 Jan 2013.

[4]     "Artificial Intelligence." *Artificial Intelligence*. Wikipedia, 3 1 2013. Web. 5 Jan 2013.

[5]     "Problem_solving" *Problem Solving*. Wikipedia, 23 12 2013. Web. 25 Dec 2013.

[6]     "Minimax." *Minimax*. Wikipedia, 23 1 2013. Web. 2 Jan 2013.

[7]     "Astar." *Astar*. Wikipedia, 23 1 2013. Web. 15 Jan 2013.

[8]     "Resolution_in_first_order_logic." *Resoultion*. Wikipedia, 23 1 2013. Web. 16 Dec 2013.

[9]     "Unification." *Unification*. Wikipedia, 23 1 2013. Web. 2 Dec 2013.

[10]    "Probabilistic_logic." *Probabilistic_logic*. Wikipedia, 23 1 2013. Web. 25 Jan 2013.

[11]    "Reinforcement_learning." *Reinforcement_learning*. Wikipedia, 23 1 2013. Web. 25 Jan 2013.

[12]    "Alpha_beta_Pruning." *Alpha_beta_pruning*. Wikipedia, 6 8 2012. Web. 15 Aug 2012.

[13]    "Knowledge_representation_and_reasoning."*Knowledge_representation_and_reasoning*. Wikipedia, 23 1 2013. Web. 25 Jan 2013.

[14]    "Partially_observable_Markov_decision_process."*Partially_observable_Markov_decision_process*. Wikipedia, 23 1 2013. Web. 13 Jan 2013.

[15]   Goodman, Len; Lauschke, Andreas; and Weisstein, Eric W. "Dijkstra's Algorithm." From *MathWorld*--A Wolfram WebSource

[16]   Eranki, Rajiv. "Pathfinding using A*." *Searching using A*.* MIT, n.d. Web. 11 Jan 2013.

[17]   Poole, David. "Uniformed Search Strategies." *Foundations of Computational Intelligence*. N.p.. Web. 25 Jan 2013.

[18]   Quevedo, J. Ubaldo. (2006, February). Heuristic Search. *Introduction to Artificial Intelligence*. University of Wisconsin, Parkside

[19]   Reeves, Colin. Genetic Algorithms, *MetaHeuristicas*. School of Mathematical and Information Sciences, Coventry University, Coventry, UK

[20]   Sinapova, Lydia. Planning, *Artificial Intelligence*. Simpson University, CA, US

[21]   Sowa, John F. "Semantic Networks." *Semantic Networks*. 6 3 2006 Web. 15 Jan 2013.

[22]   Alechina, Natasha. "Planning and Search" *Partial Order Planning*. 11 12 2012. Web. 23 Jan 2013

[23]   Nau, Dana. "First Order Logic" *Intro to AI*. Web. UMD MD 18 Jan 2013

[24]   Murphy, Kevin. "A Brief Introduction to Graphical Models and Bayesian Networks" 1998. Web.

[25]   Ingargiola, Giorgio. "Constraint Satisfaction Problems" *Introduction to Artficial Intelligence*. Web. 3 Jan 2013

[26]   Tomasi, Carlo. (September, 2011). "Image Formation" *Computer Vision*. Duke University

[27]   L. Baum et. al. A maximization technique occuring in the statistical analysis of probablistic functions of markov chains. Annals of Mathematical Statistics, 41:164–171, 1970.

[28]   Bartneck, C., Lyons, M.J., & Saerbeck, M. (2008). The Relationship Between Emotion Models and Artificial Intelligence. Proceedings of the Workshop on The Role Of Emotiono In Adaptive Behaviour And Cognitive Robotics

in affiliation with the 10th International Conference on Simulation of Adaptive Behavior: From Animals to Animates (SAB 2008), Osaka.

[29] Ortony, A., Clore, G., Collins, A. 1988. *The cognitive structure of emotions.* Cambridge: Cambridge University

[30] University Press.P.255 in Anderson, J.R. (2005). Cognitive psychology and its implications (6th edition). New York: Worth.

[31] From p.379 in Willingham, D.T. (2007). Cognition: The thinking animal (3rd edition). New Jersey: Pearson.

[32] Mayer, R. E. (1992). Thinking, problem solving, cognition. (2nd Ed.). New York: W. H. Freeman and Company.

[33] Kahneman D., Slovic P., and Tversky, A. (Eds.) (1982) Judgment Under Uncertainty: Heuristics and Biases. New York: Cambridge University Press

[34] H. Cohen & C. Lefebvre, eds., Handbook of Categorization in Cognitive Science, Elsevier, 2006, pp. 141-163.

[35] The cognitive niche: Coevolution of intelligence, sociality, and language (PNAS)Steven Pinker

[36] Lefebvre C., & H. Cohen (Eds.) (2005) Handbook on Categorization. Elsevier

[37] George A. Miller, "The Magic Number Seven, Plus or Minus Two: Some Limits on our Capacity for Processing Information", Psychological Review, Vol. 63, No. 2, 1956.

[38] McLeod, S. A. (2007). Visual Perception

[39] Coltheart, M. Modularity and Cognition. Trends in Cognitive Sciences, 3:3, 1999.

[40] R. Schalkoff, Artificial Neural Networks, Toronto, ON: the McGraw-Hill Companies, Inc., 1997.

[41] K. P. Murphy, "Dynamic Bayesian Networks: Representation, Inference and Learning", PhD thesis. UC Berkeley, Computer Science Division, July 2002.

[42] Kjrulff, U. (1992), A computational scheme for reasoning in dynamic probabilistic networks , Proceedings of the Eighth

Conference on Uncertainty in Artificial Intelligence, 121-129, Morgan Kaufmann, San Francisco.

[43]    "Robotics." *Robotics.* Wikipedia, 3 1 2013. Web. 5 Jan 2013.

[44]    "Artificial Intelligence." *Artificial Intelligence (Video Games).* Wikipedia, 3 1 2013. Web. 5 Jan 2013.

[45]    "Robot." *Robot.* Wikipedia, 3 1 2013. Web. 5 Jan 2013.

[46]    "Robot_mapping." *Robot_mapping.* Wikipedia, 3 1 2013. Web. 5 Jan 2013.

[47]    *Robot Motion Planning*, Jean-Claude Latombe, 1991, Kluwer Academic Publishers

[48]    "Behavior-based Robotics Perception." . Idaho National Laboratory. Web. 12 Nov 2012.

[49]    "Monte Carlo Localization: Efficient Position Estimation of Mobile Robots," by Dieter Fox, Wolfram Burgard, Frank Dellaert, and Sebastian Thrun, *Proceeding of the 1999 National Conference on Artificial Intelligence (AAAI)*

[50]    Choset, Lynch, Hutchinson, Kantor, Burgard, Kavraki, Thrun. (September 14, 2007). *"Principles of Robot motion"* Biorobotics. Carnghie Mellon University

[51]    Blackmore, Stout, Wand, Runov. "ROBOTIC AGRICULTURE–THE FUTURE OF AGRICULTURAL MECHANISATION?" 5th European Conference on Precision Agriculture, Uppsala, Sweden, 9-12th June 2005, Blackmore.

[52]    "The PID Algorithms" *Straight Line Control.* Web. 22 Nov 2012

[53]    Zheng, Beckey, Sanderson. "Chapter 6" ROBOTICS FOR BIOLOGICAL AND MEDICAL APPLICATIONS

[54]    Nelson, B. and Y. Zheng. 2004. *Status of robotics in the U.S.: Bio/Pharmaceutical.* NSF Workshop on status of robotics in the United States, Arlington, Virginia, July 21–22

# INDEX